D0410981

Particulate Matter and Aquatic Contaminants

Editor

Salem S. Rao, Ph.D.

National Water Research Institute
Canada Centre for Inland Waters
Burlington, Ontario, Canada

Library of Congress Cataloging-in-Publication Data

Particulate matter and aquatic contaminants / edited by Salem S. Rao.
p. cm.
Includes bibliographical references and index.
ISBN 0-87371-678-7
1. Water—Pollution. 2. Particles—Environmental aspects.
3. Environmental chemistry. I. Rao., S. S. (Salem S.), 1934– .
TD425.P37 1993
628.1′68—dc20 92-38378
CIP

Direct all inquiries to CRC Press, Inc., 2000 Corporate Blvd., N.W., Boca Raton, Florida 33431.

PRINTED IN THE UNITED STATES OF AMERICA
1 2 3 4 5 6 7 8 9 0
Printed on acid-free paper

PREFACE

The impetus behind the compilation of this volume is the ever-increasing awareness of the importance of microbial-contaminant interactions in aquatic environments. The role of suspended primary particles, such as clay or silt, in the adsorption and transportation of contaminants in aquatic systems is well recognized; however, there are still many unknowns regarding the importance and relevance of suspended particulates in the concentration and transport of contaminants. It has been surmised that a fuller understanding of both adsorption and transportation processes would help in the development of contaminant transport models and management strategies.

The eight chapters in this volume will enable the reader to gain a better understanding of suspended particulate-contaminant interactions and some of the biological, microbiological, and ecotoxicological principles associated with contaminant adsorption and transportation processes. Information and techniques that are at the leading edge of biological-contaminant transport research will be found in this volume as well as information addressing a number of toxic contaminant issues that are of global concern.

The contents of this volume will be of special interest to environmental researchers, managers of aquatic contaminant research, and modelers involved in contaminant transport.

Salem S. Rao

EDITOR

Salem S. Rao, Ph.D., RSM (CCM), is a Research Scientist with the National Water Research Institute, Canada Centre for Inland Waters, Burlington, Ontario, Canada, working in the area of environmental microbiology and ecotoxicology. He is an Adjunct Professor at Brock University, St. Catharines, Ontario and holds an associate position at the University of Toronto. He is the Editor/Author of the book ACID STRESS AND AQUATIC MICROBIAL INTERACTIONS (CRC Press) and serves on the editorial boards of the journals, *Environmental Pollution* (Elsevier Publication) and *International Journal of the Great Lakes Research.*

Dr. Rao has worked in Aquatic and Environmental Microbiology and Ecotoxicology for over 20 years. During his professional career he has published over 100 research papers and reports on such diverse topics as bacterial populations and processes in the Great Lakes and acid-stressed lakes, psychrophilic bacterial activity, bacterial-phosphate kinetics, and sulphur cycle dynamics in acidified lakes.

Dr. Rao earned a B.Sc. (Hons) degree in Zoology and Chemistry in 1956 and a Ph.D. in 1964 in Zoology and Microbiology. He is a registered microbiologist with the Canadian College of Microbiologists and has been nominated to The Royal Society of Canada.

CONTRIBUTORS

Terrance J. Beveridge, Ph.D.
Department of Microbiology
College of Biological Science
University of Guelph
Guelph, Ontario, Canada

Philip J. Bremer, Ph.D.
Seafood Research Laboratory
New Zealand Institute for
Crop & Food Research
Port Nelson, New Zealand

Shirland A. Daniels, Ph.D.
National Water Research Institute
Canada Centre for Inland Waters
Burlington, Ontario, Canada

Bernard J. Dutka, M.Sc.
National Water Research Institute
Canada Centre for Inland Waters
Burlington, Ontario, Canada

Gill G. Geesey, Ph.D.
Center for Interfacial Microbial
Process Engineering
Montana State University
Bozeman, Montana

William L. Kutas, B.A.
Ministry of the Environment
London, Ontario, Canada

Gary G. Leppard, Ph.D.
Rivers Research Branch
National Water Research Institute
Environment Canada
Burlington, Ontario, Canada
and
Department of Biology
McMaster University
Hamilton, Ontario, Canada

Dawn E. McLean
Ministry of the Environment
London, Ontario, Canada

Sandra M. Meissner
Ministry of the Environment
London, Ontario, Canada

Aaron L. Mills, Ph.D.
Laboratory of Microbial Ecology
Department of Environmental
Sciences
University of Virginia
Charlottesville, Virginia

Garry A. Palmateer, M.Sc.
Ministry of the Environment
London, Ontario, Canada

Salem S. Rao, Ph.D.
National Water Research Institute
Canada Centre for Inland Waters
Burlington, Ontario, Canada

James E. Saiers, M.Sc.
Laboratory of Microbial Ecology
Department of Environmental
Sciences
University of Virginia
Charlottesville, Virginia

Colin M. Taylor, B.Sc.
National Water Research Institute
Canada Centre for Inland Waters
Burlington, Ontario, Canada

Joel B. Thompson, Ph.D.
Marine Science Department
Eckerd College
St. Petersburg, Florida

Lesley A. Warren, B.Sc.
Department of Zoology
University of Toronto
Toronto, Ontario, Canada

Ann P. Zimmerman, Ph.D.
Department of Zoology
University of Toronto
Toronto, Ontario, Canada

ACKNOWLEDGEMENTS

The Editor expresses his sincere thanks to all contributors. He is especially appreciative of the thoughtful suggestions, sustained support, and valued encouragement of Mr. B. J. Dutka who has directly contributed to the timely development of this environmentally relevant volume.

CONTENTS

CHAPTER 1

Suspended Particulate/Bacterial Interaction in Agricultural Drains

Garry A. Palmateer, Dawn E. McLean, William L. Kutas, and Sandra M. Meissner

TABLE OF CONTENTS

0-87371-678-7/93/$0.00+$.50

I. INTRODUCTION

The pleasure of utilizing the bathing beaches situated on the Great Lakes, as well as the inland beaches, in the province of Ontario, Canada has been greatly curtailed during the past eight years because of excessively high levels of fecal bacteria. Detailed investigations into the reasons for bacterial contamination of recreational waters have revealed a set of circumstances that appears to be common to many of the beaches. The sources of fecal bacteria are often associated with agricultural runoff resulting from mismanagement of manure.[1] In more developed areas of the province, urban wastes, comprised of sewage treatment plant wastes, storm sewer discharges, and specific industrial wastes, constitute the sources of fecal-associated bacteria in the majority of cases.

In evaluating how the bacteria affect water quality of bathing beaches, it was evident that the rivers and streams carrying bacteria from their points of entry into the receiving streams were contaminated with fecal-associated bacteria at certain times and not at others. The routine sampling of the rivers impacting on the beaches indicated that the total and fecal bacterial loadings were actually quite infrequent. The bacterial quality of the beaches themselves fluctuated greatly, making management of the beaches for recreational activities very difficult. Specifically, fecal coliform and *Escherichia coli* levels could exceed the bathing beach guideline for 100 cells per 100 ml of water by one order of magnitude on one day, while the next day the levels could be less than 10 cells per 100 ml of sample. It was observed that coincident with high bacteria levels were high turbidity levels. It appeared that when the waters were rough, with the wave height exceeding 60 cm, the bacterial levels were excessive, as were the particulate levels in the water. The impact of high turbidity levels in rivers on the beaches of Lake Huron is displayed in Plates 1 and 2. Conversely, when the beach waters were calm, bacterial levels were significantly lower (i.e., below the standard) and the waters were clear (free of particulates).[2]

The association of bacteria with soil particulates was easily observed with farm drainage, whether the runoff was surface or subsurface. Subsurface drainage from tiled fields, which had manure applied, contained bacteria and

PLATE 1. Turbidity plume of the Old Ausable River discharging to Lake Huron.

PLATE 2. Turbidity plume of the Ausable River discharging to Lake Huron.

soil particulates. Recent studies conducted by Dean et al.[3] have demonstrated rapid infiltration of liquid manure applied to fields through to the underlying tiles. This occurred as liquid manure, containing levels of fecal coliforms and *E. coli* at 10^6 cells per 100 ml of manure, penetrated macropores in the surface-soil horizons. The drainage exiting the field tile into agricultural drains, which discharge into rivers impacting bathing beaches, contained levels of fecal bacteria and soil particulates ranging from 10^3 to 10^5 cells per 100 ml. The association of these bacteria with the soil, particulates has been, until recently, only speculated. The lengthy survival of fecal bacteria in soil, however, has been known for some years,[4,5] and is now being related to the transport of total and fecal bacteria in farm drainage. The transport of soil particulates resulting from erosion and runoff from farmland has been studied in detail by sedimentologists.[6] The transport of chemical constituents of soil and sediment particulates, including pesticides and metals, has been described. It is realized that soils containing high concentrations of clay (montmorillonite with a high cation exchange capacity) and organic matter have the ability to sorb significant quantities of chemical components. Some research has shown that viruses, such as bacteriophage, also tend to be attracted to these same constituents of soil because of surface charges on the bacteriophage and the clay and organic matter.[7]

Studies on the transport of suspended sediments have demonstrated that particulates that are routinely suspended in the water column tend to range from 0.1 to 200 μm in diameter. Particulates above this size may be suspended during flood periods, but average stream flows of rivers and streams do not have the energy to resuspend these particulates.[6] Where stream-flow energy is sufficient, fine sands (20 to 200 μm), silts (2 to 20 μm), and clays of less than 2 μm in diameter, all are suspended in the stream, which results in their transportation downstream.

The clay faction, which is an ideal sorbent, is usually the most erodible, as it is the finest soil component. In Ontario, suspended particulates from agricultural areas are comprised of greater than 60% clay, with the contents ranging from 59% to 98% clay.[6] This represents a significant amount of clay that can be suspended in streams, and an increase from one to four times the amount of clay that was originally in the soil prior to its erosion into the stream. It is this clay component of soil that acts as a major transporter of pollutants downstream.

Preliminary studies have shown that various size fractions of suspended particulates in streams also conduct the transport of bacteria downstream. It was shown that certain size fractions of suspended particulates tended to be highly colonized by bacteria at specific times of the year. Specifically, at the headwaters of the Desjardine Drain during the summer months, particulates become colonized with bacteria at a concentration of one bacterium per 4 μm^2 of surface area (or 2.1×10^5 bacteria per mm^2 of particulate surface area). Characterization of particulates found at the discharge of the drain to those detected at a beach 18 km from the headwaters show a similar degree of colonization. One bacterium was found for every 1.6 μm^2 of surface area. The mean concentration was found to be 6.5×10^5 bacteria per mm^2. The particulates colonized ranged from 2 to 5 μm in diameter.[8]

During the late fall, the examination of the particulates in the water showed a significant change in condition. At the headwaters of this agricultural drain, one bacterium per 2.2×10^3 μm^2 of particulate surface area was detected. This equates to 450 bacteria per mm^2 surface area. The Grand Bend Beach, which is impacted by the Desjardine Drain, also exhibited a decrease in bacterial colonization of suspended particulates as 1 bacterium per 5.1×10^3 μm^2 surface area was observed, or 195 bacteria per mm^2. Most bacteria were observed on particulates ranging in diameter from 2 to 5 μm. The average percent viability of the sorbed bacteria declined from 50 in the summer to 17 in the fall.

These observations provide some evidence as to the rate of colonization of suspended particulates in an agricultural drain that impacts on a bathing beach of the Great Lakes.

The attachment processes of bacteria to soil particulates have been described by Marshall.[9] Surface charges on the particulates and the bacteria, which have been classified as weak van der Waals' forces, and which, in turn, exceed other repulsive forces, are sufficient to cause a net attractive effect. The resulting adsorption is termed reversible sorption. Although the bacteria and the particulate surface do not physically attach, there is a definite concentrating effect of the amount of bacteria on the surface of the particulates. It is speculated that fecal-associated bacteria may initially become sorbed to suspended particulates in this fashion. This type of sorption is, however, termed reversible because the bacteria may be readily desorbed from the particulate surface by shearing forces, caused by the turbulence such as that associated with wave action at a beach or a riffle in a river or stream. Three basic orientations of bacteria to particulates are depicted in Figure 1.

Permanent sorption involves direct contact of the microorganism with the surface of the particulate by the anchoring of the bacterium, for example, by pili, fimbriae, and flagella to a specific sorption site. Mucopolysaccharide, excreted by the microorganisms, has the effect of cementing the organism to the sorption surface.[10] These two basic concepts are illustrated in Figure 2.

In summary, the attachment of microorganisms, such as bacteria, to suspended particulates in streams tends to support lengthy bacterial survival because the nutrients, which are already sorbed to portions of the particulates, are readily available for bacterial uptake. Paerl[11] found the levels of nutrients to be 10 to 100 times higher on the particulate surface than in the surrounding water column.

Once bacteria and other microorganisms are attached to particulates, other benefits are accrued. For example, clays provide a buffering capacity of acidic aqueous conditions, thereby providing a more suitable environment for acid-sensitive bacteria.[12,13] Clays also affect microorganism survival by providing protection from UV and X-irradiation, desiccation, antibiotics, and predator-prey interactions.[10] The net result of sorption of bacteria to organic and inorganic sorbents like clay is the increase in survival from a few days, in the overlying waters, to several months on the particulates in the sediment.[14]

Finally, concurrent with the survival on bottom sediments and soil particulates, are the long distances (greater than 10 km) that bacteria and other microorganisms, such as fungi, algae, and viruses, may travel once sorbed to readily suspended and transportable particulates (less than 100 μm in diameter).

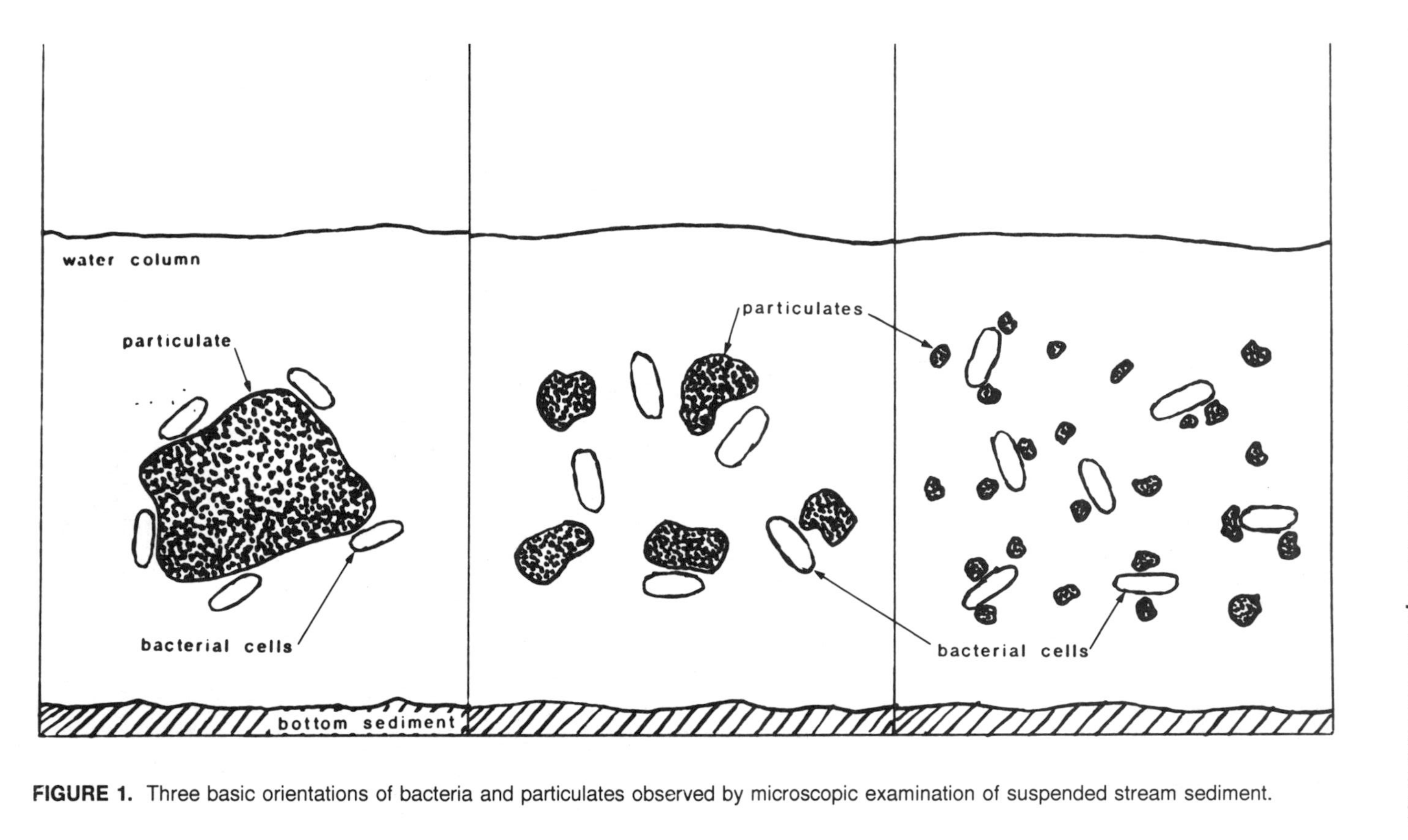

FIGURE 1. Three basic orientations of bacteria and particulates observed by microscopic examination of suspended stream sediment.

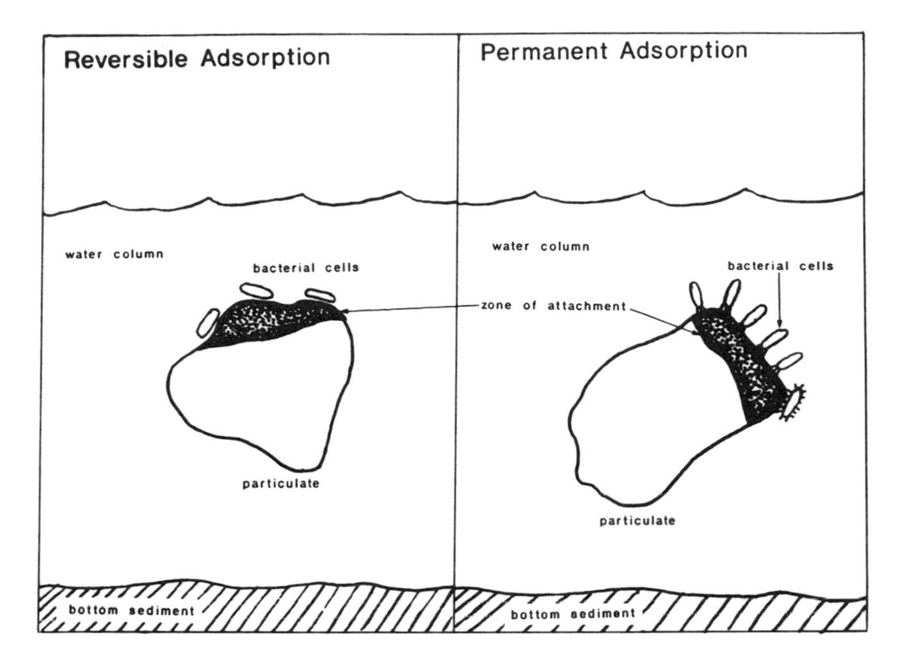

FIGURE 2. Illustration of bacteria and particulate orientation describing reversible and permanent adsorption.

The focus of this study was to investigate the degree of colonization of suspended particulates by total viable bacteria and *Salmonella*. Three different agricultural drains were studied during the summer, fall and spring seasons, using the latest techniques in epifluorescence and immunofluorescence microscopy for detecting viable bacteria on sediment particulates.[15-17]

In addition, a bacterial transport study was conducted where sediment-bound *E. coli*, labeled with nalidixic acid resistance (NAL), were introduced into the stream and were monitored to determine the distance transported on indigenous sediments of the agricultural drains. These results were then compared to similar studies done with nonparticulate-bound bacteria.

II. METHODS AND MATERIALS

A. Sampling Sites

The three agricultural drains were chosen because they represented different degrees of impact by rural land-use activities, and they each affected the quality of their respective beach waters.

The first location was the Arthur Vanatter Drain near the village of Kintore in Oxford County, as shown in Figure 3. The upstream location was

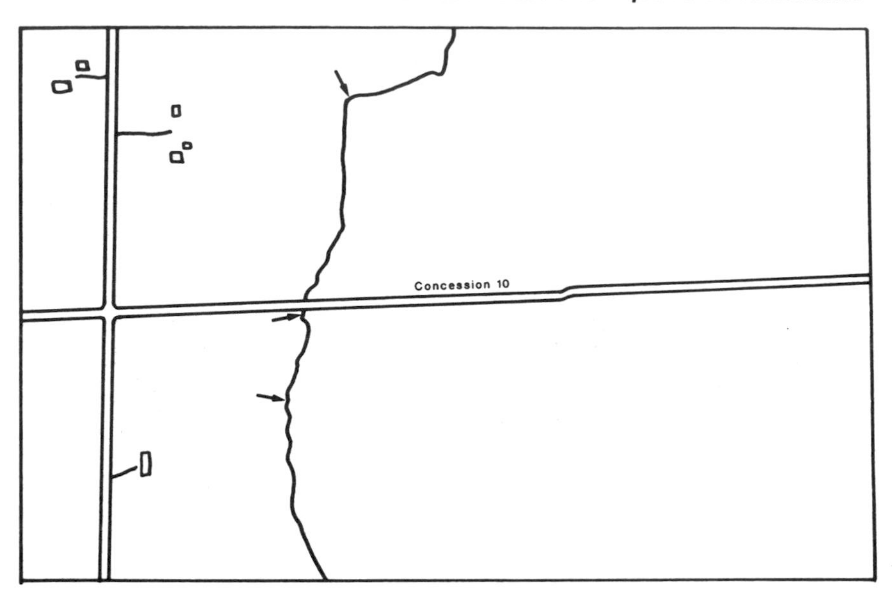

FIGURE 3. Map of the Arthur Vanatter Drain study site.

approximately 300 m north of the main site, a culvert on Concession Road 10. The downstream site was located 200 m south of the concession road. Corn was planted along the edge of the drain (within 4 m), and manure was applied as liquid swine waste only on the south side of the concession road, usually in the spring. The drain discharges to the middle branch of the Thames River.

The second sampling location, along the Central School Drain near Shakespeare in Perth County, was also comprised of three study sites, as shown in Figure 4. The sites were accessed by Concession Road 3. The upstream site, 100 m north of the road, was bounded by pasture on which cattle were able to water in the drain. The main site location, at the road culvert, was also impacted by the cattle, but in addition, a field drainage tile discharged into the drain that contained field-applied manure and milkhouse wastes. The downstream site was located 150 m from the road. The drain discharges to the Avon River, approximately 1 km to the south.

The third sampling location was located on the Desjardine Drain near Grand Bend in Huron County, as shown in Figure 5. The upstream site was situated 800 m east of the Playhouse Road. This site received farm wastes from upstream farming operations, but was located in a marshy woodlot. The main site, to the west, was located at the access road. The surrounding land activities include beef farming and six or seven homes, all of which have septic-tank tile beds adjacent to the drain. The downstream site was located 800 m west, and was also impacted by beef cattle and septage from rural septic tanks. The Desjardine Drain discharges into the Old Ausable River which, in turn, discharges to Lake Huron at Grand Bend Beach.

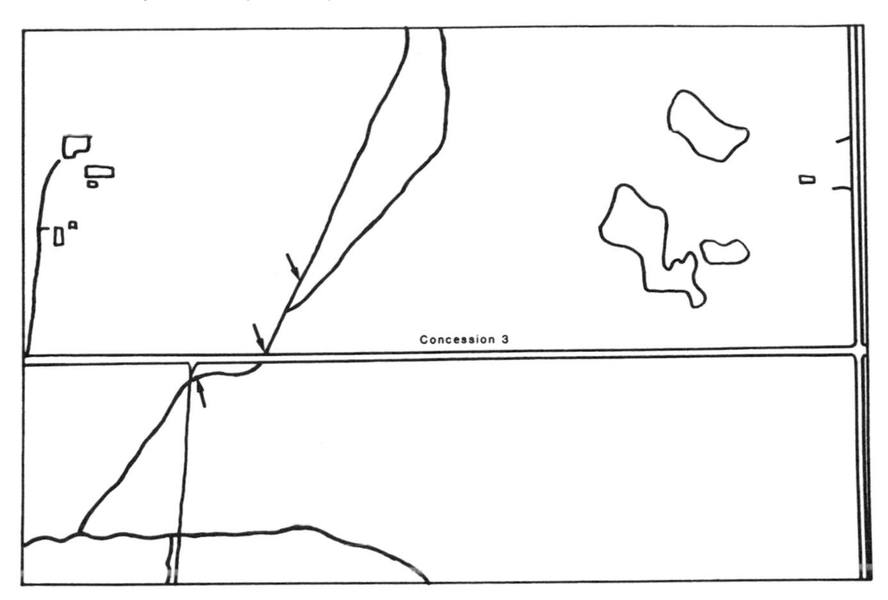

FIGURE 4. Map of the Central School Drain study site.

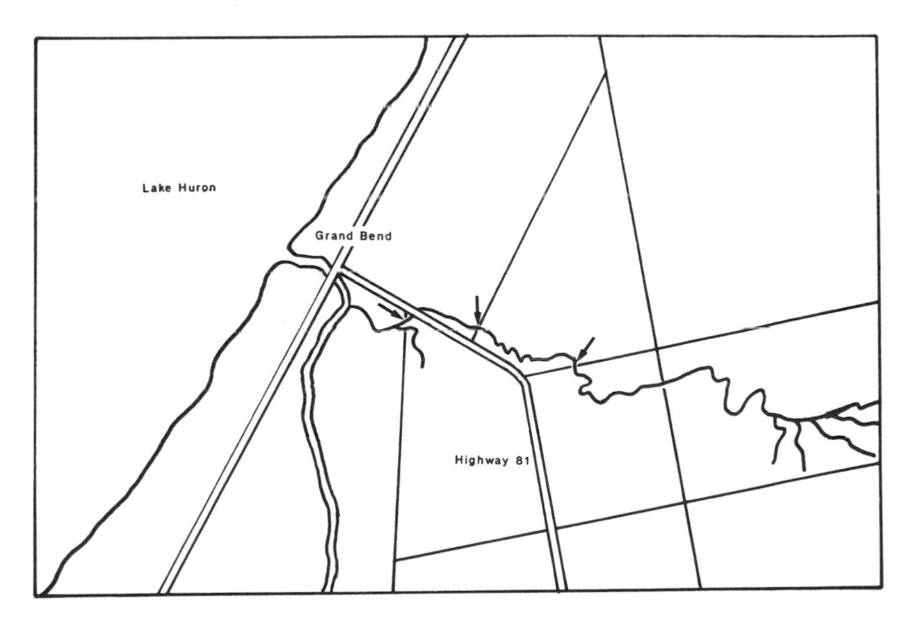

FIGURE 5. Map of the Desjardine Drain study site.

B. Sampling Protocol

The study commenced in the summer of 1991, with the manual sampling of the three sites in each of the three locations described above.

The samples for suspended particulates were taken in sterile, 500-ml widemouth plastic bottles placed in the main stream flow. The bottle design provided for rapid filling of suspended particulates with minimal shearing of the delicate aggregates.

Samples for the standard bacterial indicator parameters, *Escherichia coli* and fecal streptococci, were taken in sterile, glass, milk-dilution bottles.

The bacterial transport study involved sampling, in duplicate, using conventional bacteriological sterile, glass, milk dilution-bottles. Composite samples were taken using specifically constructed cotton swabs.

C. Particulate Characterization Procedure

The sizing and counting of suspended particulates was conducted using an Elzone 180 XY Particle Analyzer®, manufactured by Particle Data Inc., Elmhurst, IL. Preparation for analyses included filter screening each sample with 212-μm mesh. The particulates known to be transported in streams and rivers and known to act as sorbents for microorganisms are less than 200 μm in diameter.

The filtrate was collected and diluted in sodium hexametaphosphate (Calgon®, Benckiser Inc., Toronto, ON) according to the requirements of the instrumentation.[18] The Calgon-suspending diluent provided stabilization of the particulates from aggregation or deflocculation.

The Elzone 180 XY Particle Analyzer functions on the Coulter Principle. The system is comprised of a glass tube with an orifice at one end, which may vary in size from 12 to 900 μm, two electrodes and a vacuum pump. One electrode is inside the tube, and the other is located on the outside. The tube is placed in an aqueous medium containing electrolytes and the particles to be sized and counted. The vacuum pump draws the suspended medium through the orifice. As the particles pass by the electrodes, they cause an increase in current resistance. The change in resistance is directly proportional to the volume of the particle passing between the electrodes. The particle is then counted and sized. With this information, the Elzone 180 XY calculates particle diameter, volume, and surface area in the appropriate units of measurement.

Since the particles to be counted would vary considerably in size, different orifice tubes were used. An orifice with a diameter greater than 60% of that of the largest particles to be sized was employed. To obtain optimal resolution of particle sizes, three orifices were used with orifice diameters of 380, 95, and 24 μm. The operating size range for the 380-μm orifice tube was from 2 to 150 μm. For the 95-μm orifice tube, the operating range was 2 to 38 μm. The range for the 24-μm tube was from 2 to 10 μm. Each orifice provided a higher degree of resolution of the particle sizes for its respective operating range.

After the 380-μm orifice was used, the remainder of the sample was filtered through a 53-μm mesh size for analysis with the 95-μm orifice tube.

In preparation for the 24-μm orifice tube, the sample was finally filtered through a 12-μm mesh to remove particulates greater than the optimal size for the orifice tube.

Once preliminary analyses were conducted, the predominate particle sizes were assessed as to diameter and surface area. This information allowed for the selection of filters with pore sizes slightly smaller than each of the predominate particle-size diameter ranges. As a result, filters with 30-, 10-, 5-, and 1-µm pore sizes were chosen from the particle sizing data. The purpose of this was to selectively isolate each predominate group of particles, based on their diameter, for direct microscopic observation.

Particles were also measured microscopically to confirm the observations from the particle analyzer. This was accomplished by first filtering a sample of suspended particulates with a standard vacuum-filtration apparatus which uses 47-mm diameter filters. Initially, a 30-µm nylon mesh filter was used. The filtrate was captured and refiltered on a 10-µm pore size Nuclepore® filter (Nucleopore Corp., Pleasanton, CA). This was repeated consecutively for 5- and 1-µm pore size filters. A very slight vacuum was used in order to preserve the integrity of the particulates.

The size of the particulates was determined by using an area-calibrated graticule in an ocular of a microscope, and, through that, estimating the area covered by the particulate.

D. Bacterial Water Quality Methods

The *E. coli* and fecal streptococci water-quality parameters were measured according to the methods of Handbook of Analytical Methods for Environmental Samples (HAMES).[19]

E. Epifluorescence and Immunofluorescence Microscopic Methods

The epifluorescence and immunofluorescence microscopic methods of analyses for total viable bacteria and *Salmonella* associated with particulates (sorbed and free floating) are as follows:

1. Viability Determination

An aliquot of sample containing suspended particulates was incubated with 1 ml of 0.4% 2-*p*-iodophenyl-3-*p*-nitrophenyl tetrazolium chloride (INT), 1 ml of 0.25% yeast extract, and 1 ml of 0.001% lomefloxacin (Searle Pharmaceuticals®, Skokie, IL) per 10 ml of sample at 20°C for 4 h, in the dark. Following the 4-h incubation period, the sample was fixed with 0.6 ml of 37% formaldehyde per 10 ml of sample. Each sample was prepared as above.

2. Staining Technique

A double staining process, as described by Hoff,[20] was used to detect both total viable bacteria and *Salmonella* sp., employing a combination of 4′,6-diamidino-2-phenylindole (DAPI Sigma Chemical Co., St. Louis, MO) and fluorescein isothiocyanate fluorescent antibody (FITC-FA).

The treated samples were filtered onto 25-mm diameter Nuclepore polycarbonate black membranes with the appropriate pore size. The membranes were previously stained by soaking them in 0.0067% Sudan Black B for 24 h. Stained filters could be stored indefinitely at 4°C. Before using, the filters were washed in 0.1 *M* phosphate buffered saline. The membranes were then sterilized in the autoclave.

All solvent water used in preparing the solution described above was made with particulate-free water (Milli-Q System, Millipore Corp., Bedford, MA).

The final solution was always filter-sterilized using 0.22-μm pore size membranes.

Once the sample was filtered, 2 ml of 0.001% solution of the fluorescent stain (DAPI) was added to the membrane filter in order to stain both the free-floating bacteria and those sorbed to the particulates. DAPI was left on the membrane for 10 min in the dark, after which the DAPI was gently vacuumed from the membrane. The membrane was then rinsed three times with filter-sterilized distilled water.

Once the filter was washed free of DAPI, 0.2 ml of FITC-FA (DIFCO®, DIFCO Lab., Detroit, MI) was added to the filter and left to incubate at room temperature in the dark for 30 min. After the incubation period, 2 ml of FA buffer (DIFCO), pH 9, was added to the filter which remained in the filter funnel. The buffer then was immediately vacuumed gently from the filter. To further de-stain the filter, another 2 ml of FA buffer was added to the filter, and left for 3 min before gently vacuuming from the filter. This rinsing procedure was repeated two more times. After destaining, with FA buffer, the filter was further rinsed three times with filter-sterilized distilled water.

The filter was removed aseptically and placed on a drop of glycerol on an acid-washed microscope slide. A drop of glycerol containing phenylenediamine was placed on the slide, followed by a coverslip. The coverslip was sealed with nail polish. The slide was kept in the dark until examined. The phenylenediamine, added to the glycerol, retards fading of the fluorescence from DAPI and FITC-FA.[21]

3. *Microscopy*

The DAPI and FITC-FA-stained cells were examined using a Nikon Optiphot-2 microscope (Nippon Kogakukk, Tokyo, Japan) equipped with an episcopic fluorescence attachment EF-D using an UV-F (fluor) glycerine 100x objective. Because two stains were used to detect the respective bacteria, two different filter systems were employed in the microscope. For DAPI, the filter block was UV-1A, which used an EX365/10 excitation filter, a DM400 dichroic mirror, and a BA400 barrier filter. For the FITC-FA-stained antibody, the filter block was a B-2A. This used an EX450 ~ 490 excitation filter, a DM510 dichroic mirror, and a BA520 barrier filter. Photomicrographs were taken with a Nikon Microflex UFX-2 camera using Kodak Ektapress Gold 400 ASA film.

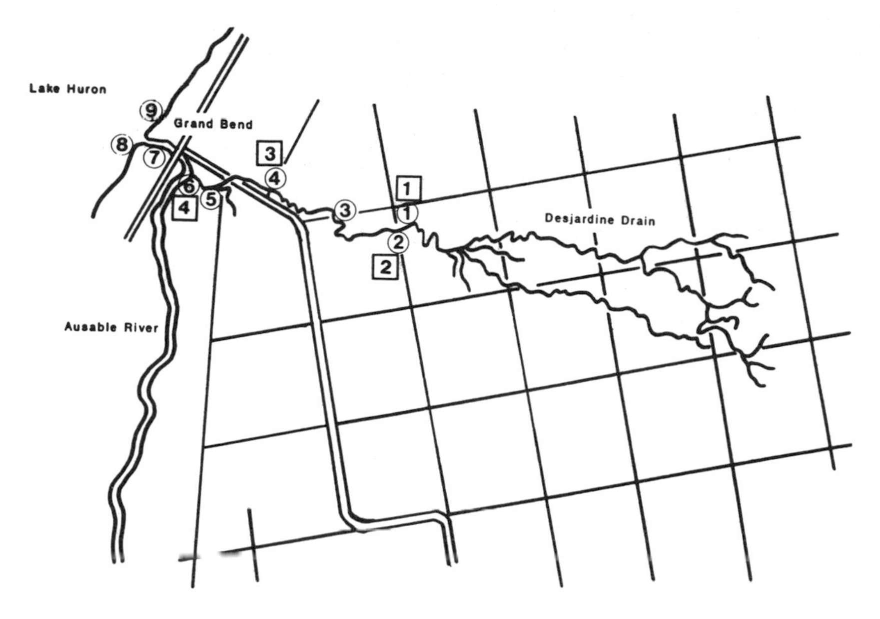

FIGURE 6. Map showing grab and swab sampling sites during the bacterial transport study. ○, sampling sites; □, swab sites.

F. Bacterial Particulate Transport Experimental Procedure

The distance the particulate-bound bacteria traveled was determined by the following method.

A 30-kg sample of the sediment was removed from a 3-cm depth at the sediment-water interface of the Desjardine Drain. The sediment was coarse-filtered to remove any leaves or other debris, and then stored at 4°C. Samples of sediment were analyzed for percent sand, silt, organic matter and clay, the pH, and the cation exchange capacity to establish the characteristics relevant to sediment sorption. The sediment was transferred to a 50-l carboy, after which 10 l of 10^7 *E. coli* nalidixic acid resistant (NAL) per ml were added. To assist in the acclimatization of the bacteria to the sediment, 50 mg of glucose per kilogram of sediment was also added. The mixture was allowed to incubate for one week at 4°C. Using the microscopic Direct Viable Cell-Count technique, a sample of the sediment was checked to assess the degree of sorption of *E. coli* NAL to the sediment, after the one-week incubation period.

The bacterial transport experiment commenced with the discharging of two 50-l carboys into the Desjardine Drain. Flow measurements made in the drain indicated the approximate travel time of the leading edge of the bacterial-particulate plume at the specific downstream sites, as shown in Figure 6.

To assist in detecting the leading edge of the plume, fluorescein dye was added to the drain at the time the bacterial-particulate mixture was discharged.

Sampling was conducted in a manner designed to detect the leading edge of the plume, the main section of the plume, and the trailing edge. In addition, to duplicate the grab samples being taken, swab samples were located at strategic sites in the drain, as indicated in Figure 6. The swabs employed were constructed as described below.

Feminine napkins were wrapped in double layers of cheese cloth. A piece of wire, 40 cm in length, was attached to each swab for the purpose of binding it to a pole placed in the stream. The swabs were wrapped in paper and sterilized in the autoclave prior to use.

Samples were collected twice a day for two days and then once a day for one week. Sampling continued once a week for another ten weeks. During the final week of study, sediment samples of the Desjardine Drain were taken at the four swab-sampling stations. The method of bacterial extraction from the sediments and the enumeration procedure were described by Palmateer et al.[22]

III. RESULTS

The bacterial water quality, with respect to fecal pollution, is shown in Table 1 for the three agricultural drains investigated during the summer, fall, and spring seasons.

The geometric means indicated degraded water quality during the low-flow period of the summer. The fall also had poor bacterial water quality. *E. coli* and fecal streptococci fluctuated greatly from season to season; however, the spring geometric means were consistently lower than those of the previous fall period for all three drains.

These data typify the bacterial water quality at the beaches where each drain was known to impact. The variation in geometric means from season to season reflected the fluctuating bacterial water quality in the water column. To understand the reasons for these fluctuations, beyond the land-use activities that affect each drain, it was necessary to further investigate the suspended particulate load that occurred.

The bottom sediment, at the sediment-water interface, was sampled and characterized, as shown in Table 2, for two seasons of the study. The Arthur Vanatter and the Central School Drains had similar results for most of the parameters. The Desjardine Drain had much less sand and considerably more silt, clay, and organic matter in the bottom sediment than did the previous two drains. The Desjardine Drain also had a significantly higher cation exchange capacity. The potential for the sediments of the Desjardine Drain to attract bacteria, as well as fungi and viruses, was much greater than that of the sediments of the other two drains.

The charge on the surface of the organic matter and the clay, as indicated by the cation exchange capacity, was shown to be partially responsible for sediment and soil sorption.[12]

Table 1. Geometric Mean Levels of E. coli and Fecal Streptococci in the Arthur Vanatter, Central School, and Desjardine Drains in Three Study Seasons

	E. coli (Geometric Mean per 100 ml)			Fecal Streptococci (Geometric Mean per 100 ml)		
Drain Site	**Summer**	**Autumn**	**Spring**	**Summer**	**Autumn**	**Spring**
Arthur Vanatter	606.3	59.1	2.5	1,771.3	155.5	7.1
Central School	2,546.2	122.8	22.5	1,326.2	1,795.9	60.2
Desjardine	401.7	699.2	15.5	391.1	1,551.3	28.0

Table 2. Characteristics of Sediments Relevant to Adsorption Processes at Three Study Locations in Summer and Spring

	Arthur Vanatter Drain		Central School Drain		Desjardin Drain	
Test	**Summer**	**Spring**	**Summer**	**Spring**	**Summer**	**Spring**
% Sand	85	77	88	28	32	69
% Silt	8	15	4	54	40	16
% Clay	7	8	8	18	28	15
% Organic matter	1.8	1.9	2.5	4.8	3.7	1.5
Cation exchange capacity	15	19	18	27	28	23
pH	8.1	7.8	7.7	7.6	7.9	7.9

A typical analysis of the suspended particulates, using the Elzone 180 XY Particle Analyzer, is shown in Figure 7. The diameters of the particulates counted and sized, for each of the three orifice tubes that were utilized, are displayed. The predominance of particulates with specific diameters can be observed. This allowed picking the filters with the pores sized slightly smaller than the diameter of each of the specific particulates, so that they could be removed from the suspended particulate population and be examined microscopically.

To visualize the relationship of the bacteria to the particulates, the following photomicrographs (Plates 3 and 4) are included.

The free-floating bacteria and the bacteria sorbed to the particulates were observed microscopically. The particulates absorb the fluorochrome to the extent that they can easily be observed in relation to the bacteria. This characteristic facilitates counting the bacteria and determining the surface area of the particulates.

A. Arthur Vanatter Drain

The levels of bacterial attachment to suspended particulates in the Arthur Vanatter Drain are displayed in Figures 8 through 11. The bacterial sorption in the summer was relatively constant in the 30- to 70-μm diameter range. The sorption decreased by approximately one logarithm in the 10- to 30-μm diameter range and again increased to 10^4 in the 5- to 10-μm diameter range. It is

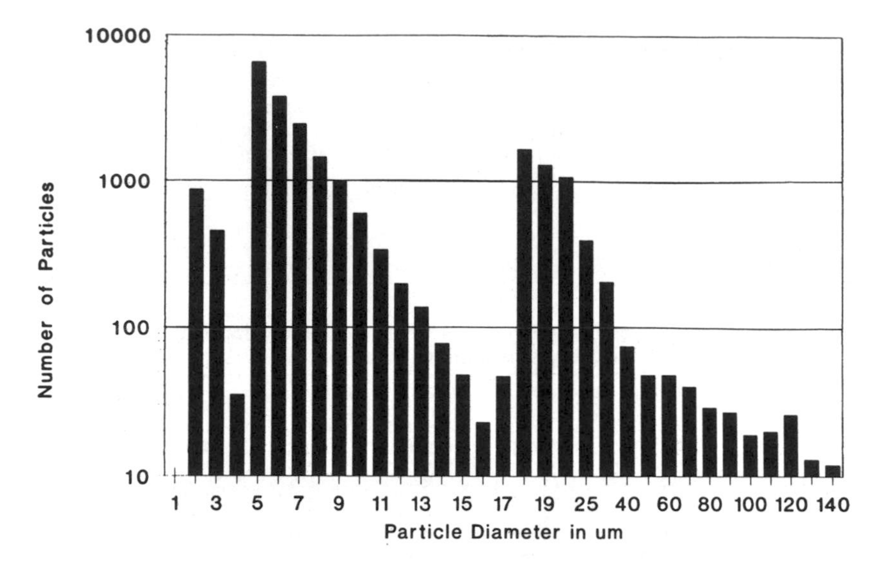

FIGURE 7. Size distribution of particulates in a typical sample of suspended sediments, based on diameter from the Elzone 180 XY.

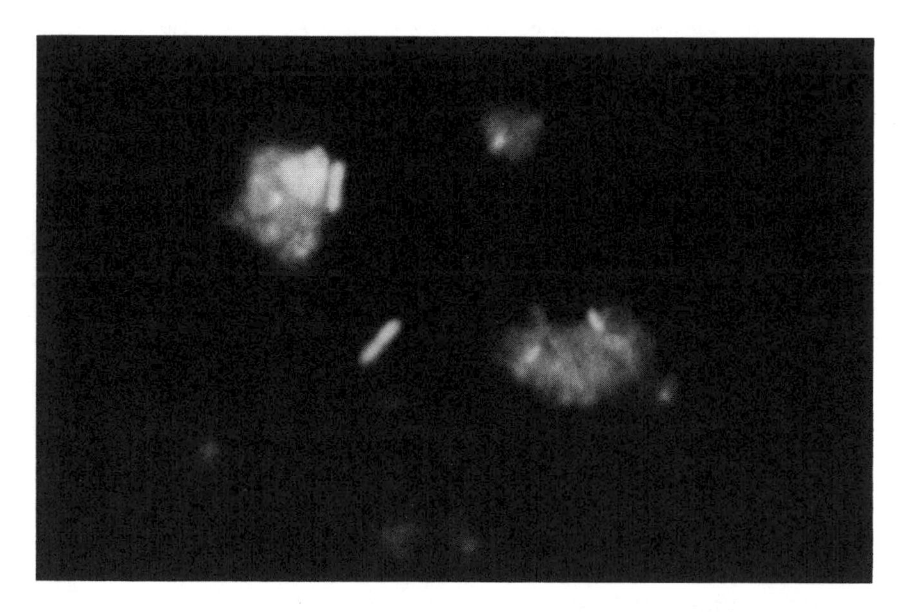

PLATE 3. Photomicrograph of bacteria sorbed to particulates and free-floating at 2200 × magnification.

PLATE 4. Photomicrograph of bacteria sorbed to particulates and free-floating at 2200 × magnification.

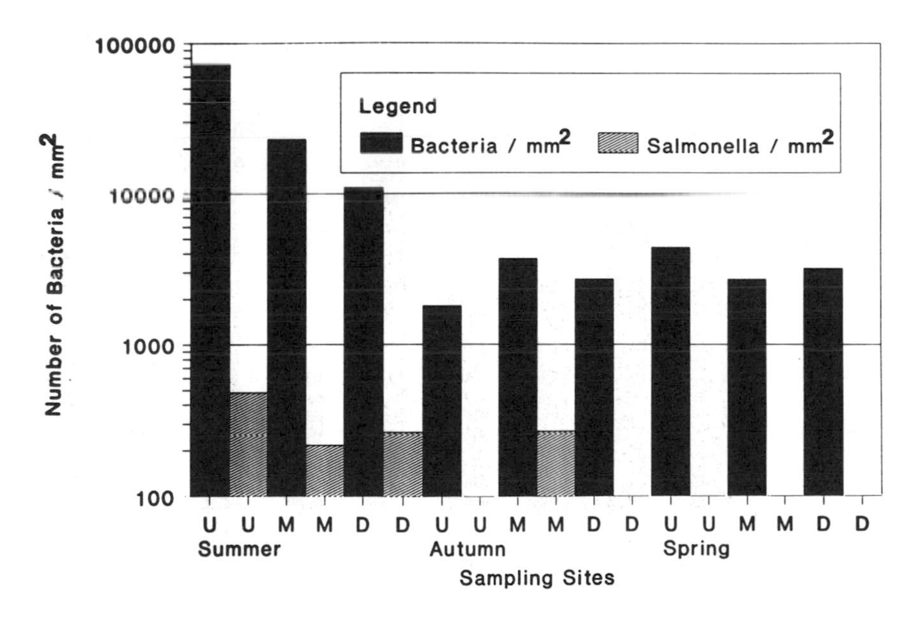

FIGURE 8. Levels of total viable bacteria and *Salmonella* per mm^2 surface area adsorbed on particles 30 to 70 μm in diameter in the Arthur Vanatter Drain. U, upstream; M, main; D, downstream.

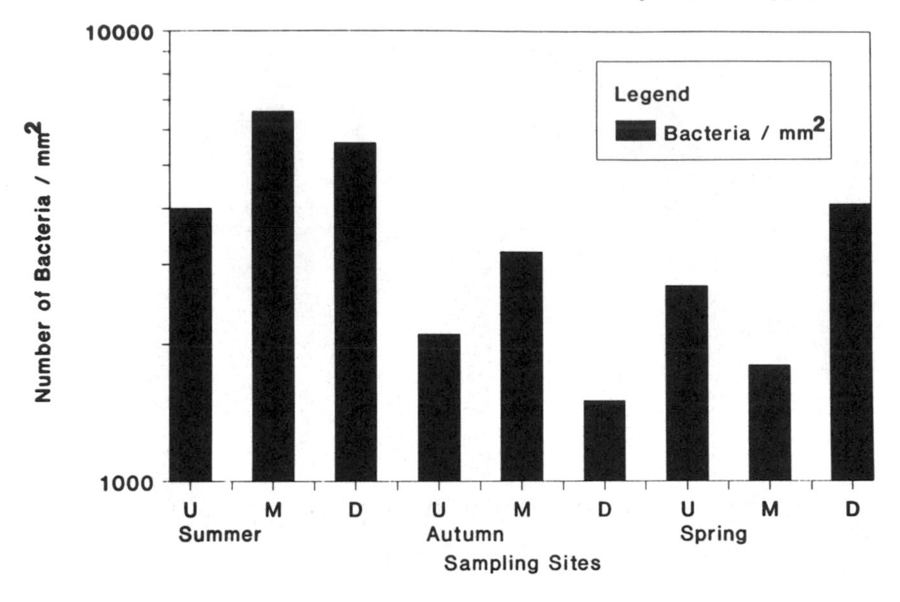

FIGURE 9. Levels of total viable bacteria per mm² surface area, adsorbed on particles 10 to 30 μm in diameter in the Arthur Vanatter Drain. U, upstream; M, main; D, downstream.

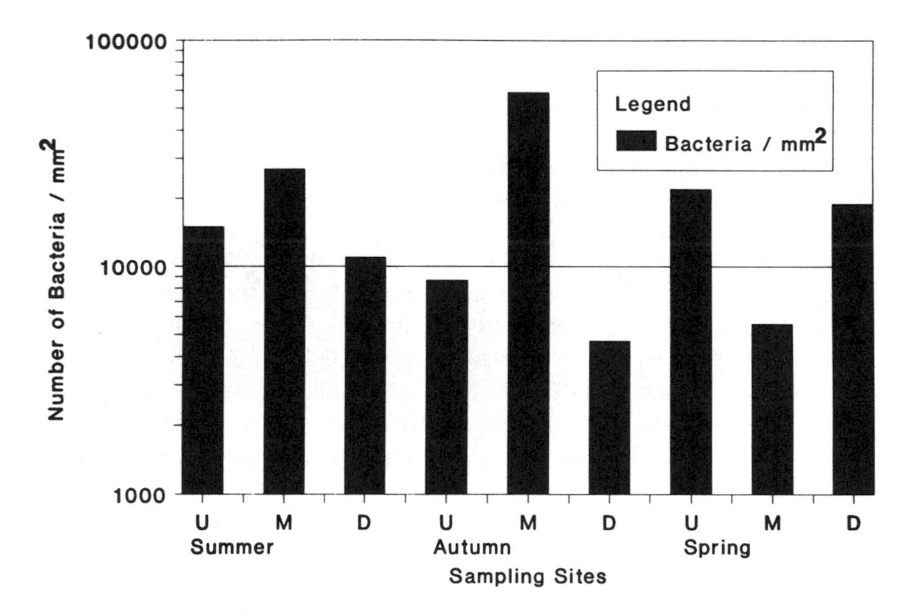

FIGURE 10. Levels of total viable bacteria per mm² surface area adsorbed on particles 5 to 10 μm in diameter in the Arthur Vanatter Drain. U, upstream; M, main; D, downstream.

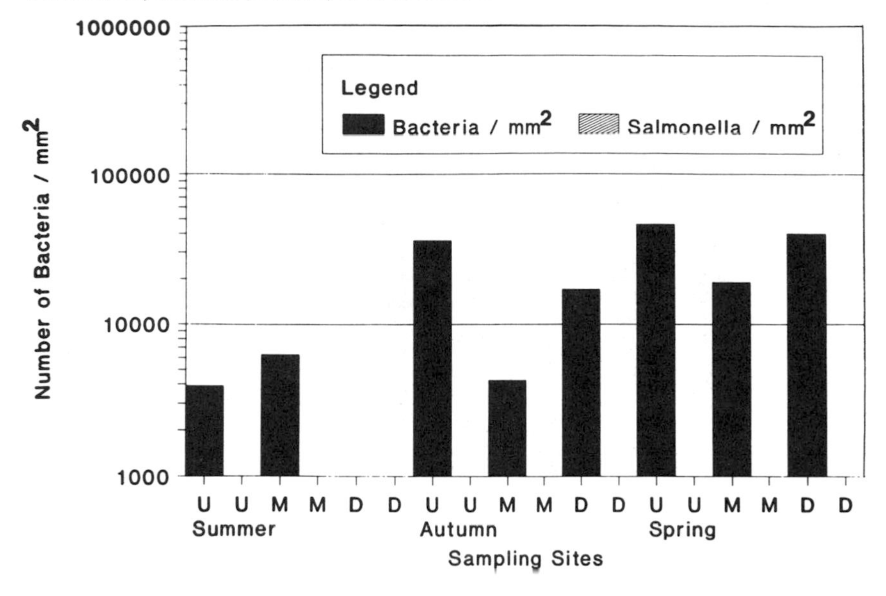

FIGURE 11. Levels of total viable bacteria and *Salmonella* per mm² surface area adsorbed on particles 1 to 5 μm in diameter in the Arthur Vanatter Drain. U, upstream; M, main; D, downstream.

notable that the level of colonization decreased to 10^3 with the 1- to 5-μm diameter range.

The autumn data, when compared to the summer data, showed bacterial decreases again for most of the size ranges.

During the spring, the level of bacterial colonization remained the same as that of the autumn level. No significant differences exist between the upstream and the downstream sites.

The free-floating bacteria associated with the 30- to 70-μm diameter particulates were only detected in the summer and autumn, as shown in Figures 12 through 15. The levels were considerably lower than the levels of bacteria sorbed to the particulates for this size range. In comparison, the 10- to 30-μm diameter size did have free-floating bacteria, which was also the case for the 5- to 10-μm and 1- to 5-μm diameter size ranges. However, the concentration of bacteria did decrease in the spring, as compared to the summer and autumn. All particulate size ranges indicated that there was an increase in the bacterial levels in the fall, which is unique to the free-floating bacteria associated with the various size ranges.

Salmonella bacteria recovery at the Arthur Vanatter Drain was observed in the summer period sorbed to particles in the 30- to 70-μm diameter range; otherwise, *Salmonella* were undetectable on any of the other particulates examined. *Salmonella* were again recovered during the autumn at 10^2 per mm² of particulate surface area. No *Salmonella* were observed in the spring.

Free-floating *Salmonella* were detected on only three occasions. The levels ranged from 10^2 to 10^3 cells per ml, as shown in Figures 12 through 15.

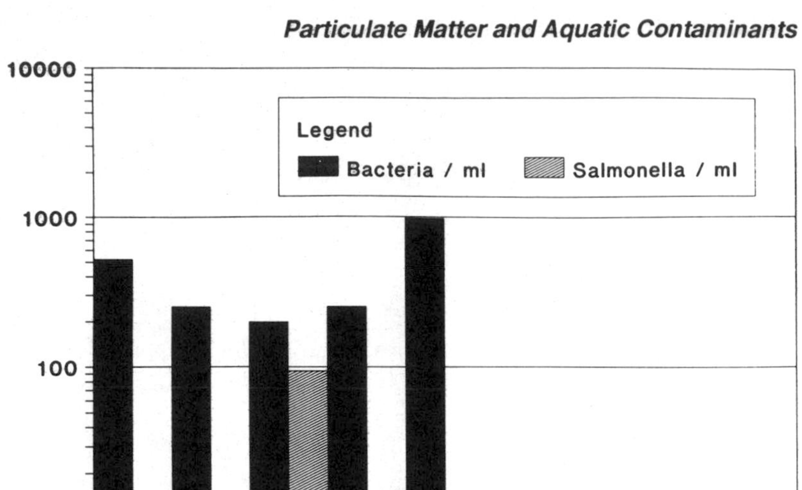

FIGURE 12. Levels of total viable bacteria and *Salmonella* that are free-floating and associated with particles 30 to 70 µm in diameter in the Arthur Vanatter Drain. U, upstream; M, main; D, downstream.

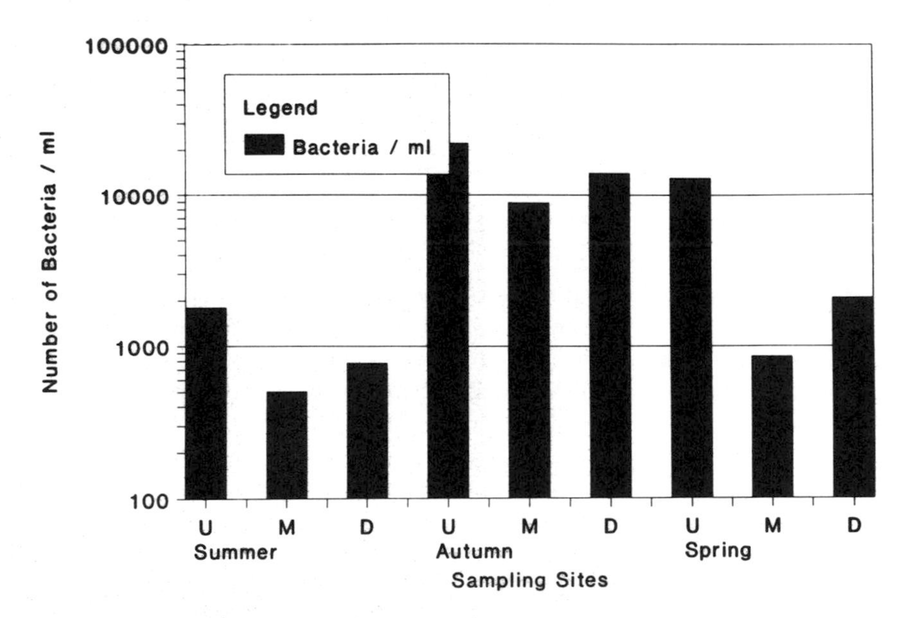

FIGURE 13. Levels of total viable bacteria that are free-floating and associated with particles 10 to 30 µm in diameter in the Arthur Vanatter Drain. U, upstream; M, main; D, downstream.

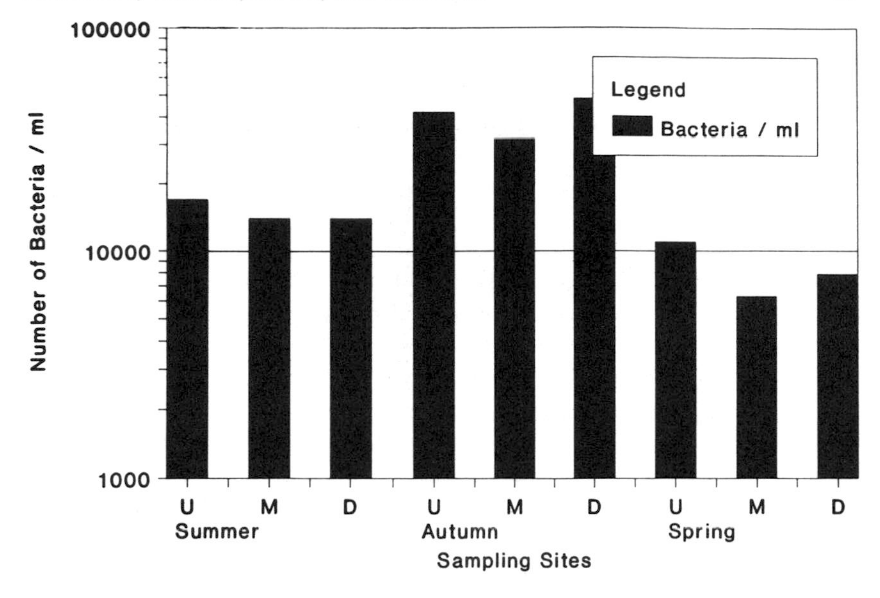

FIGURE 14. Levels of total viable bacteria that are free-floating and associated with particles 5 to 10 µm in diameter in the Arthur Vanatter Drain. U, upstream; M, main; D, downstream.

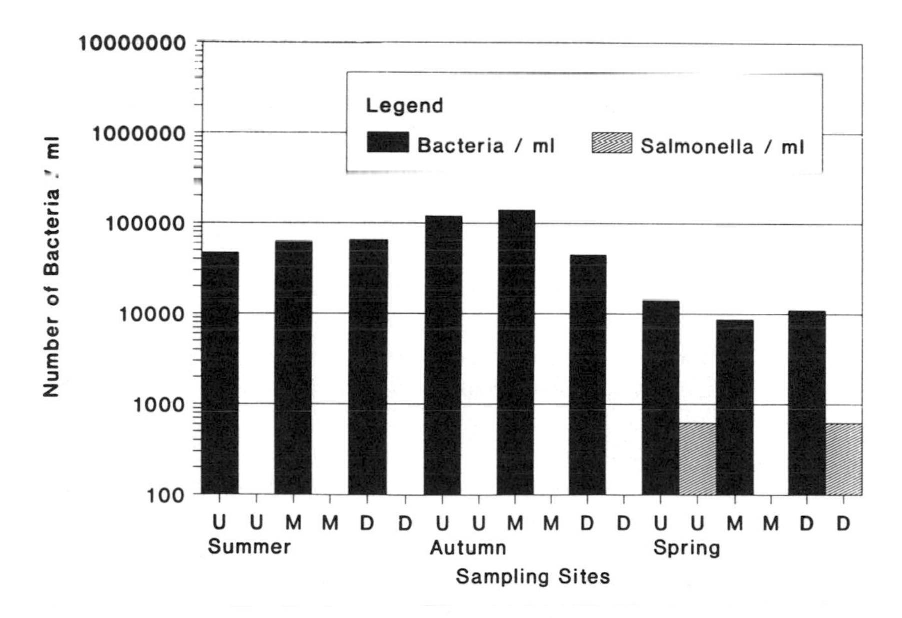

FIGURE 15. Levels of total viable bacteria and *Salmonella* that are free-floating and associated with particles 1 to 5 µm in diameter in the Arthur Vanatter Drain. U, upstream; M, main; D, downstream.

B. Central School Drain

The levels of bacteria during the summer were found to vary between 10^3 and 10^4 bacteria per mm^2 of particulate surface area for the 30- to 70-μm diameter range (Figures 16 through 19). A decrease to 10^3 bacteria per mm^2 was observed for the 10- to 30-μm diameter range. Both the 5- to 10-μm and the 1- to 5-μm diameter ranges increased to 10^4 bacteria per mm^2. During the autumn, the results show a decrease from 10^4 to 10^3 bacteria per mm^2 for the 30- to 70-μm diameter range. The other size ranges examined remained the same as they were during the summer months. The spring sampling results revealed bacterial sorption to suspended particulates to be similar to those of the summer and autumn sampling periods.

In Figures 20 through 23, fluctuating levels of free-floating bacteria can be detected. Figure 20 shows that no bacteria were recovered during the spring, that were associated with the larger particulates (30 to 70 μm in diameter). This was typical of the results.

The levels of bacteria detected free-floating with the 10- to 30-μm, 5- to 10-μm and 1- to 5-μm diameter size ranges show levels from 10^3 to 10^5 per ml. Significant levels were detected during the spring, as well as during the summer and autumn.

Salmonella were recovered from samples taken during all three seasons. They were recovered free-floating and were associated with the 1- to 5-μm diameter particulates.

Salmonella bacterial contamination of suspended particulates was observed to be minimal, as they were recovered only twice during the three periods of the study.

C. Desjardine Drain

Figures 24 through 27 display the results of total viable bacterial concentrations on the particulates for the Desjardine Drain in the format of mean numbers of sorbed bacteria per mm^2 of particulate. The free-floating bacterial concentrations are also exhibited, based on the number of bacteria per ml for each size range.

The results of the *Salmonella* determinations per mm^2 of particulate surface at the three sampling sites for each season during the Desjardine Drain study are also shown in Figures 24 through 27.

It is evident that the large particulates in both the 30- to 70-μm and the 10- to 30-μm diameter ranges were colonized by the total viable bacteria to approximately 10^4 cells per mm^2 of surface area during the summer and autumn months, with little variation. The exception occurred during the spring when the levels declined to approximately 10^3 cells per mm^2.

The *Salmonella* concentrations, as expected, were considerably lower than those of the total viable bacteria, but were still easily detectable at all of the study sites on the Desjardine Drain on particulates 30 to 70 μm in diameter.

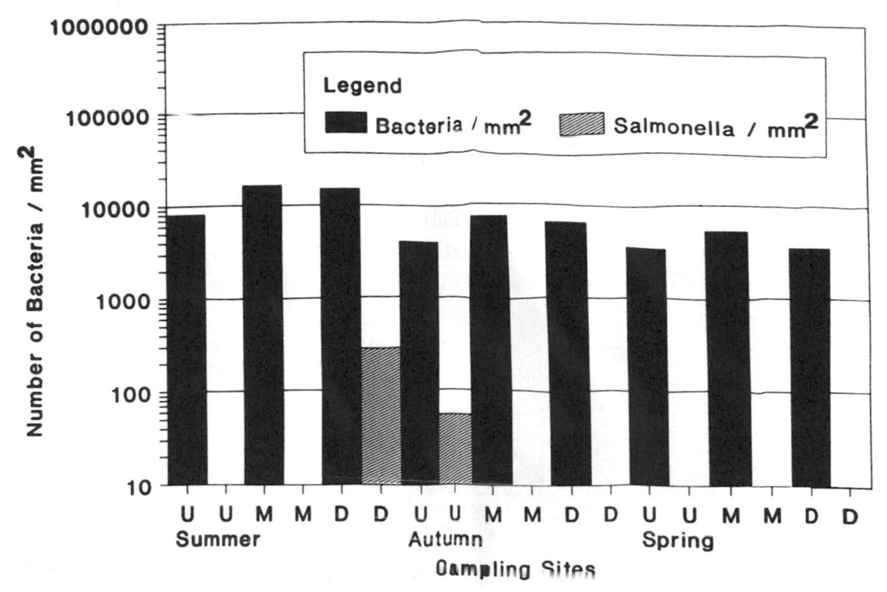

FIGURE 16. Levels of total viable bacteria and *Salmonella* per mm² surface area adsorbed on particles 30 to 70 μm in diameter in the Central School Drain. U, upstream; M, main; D, downstream.

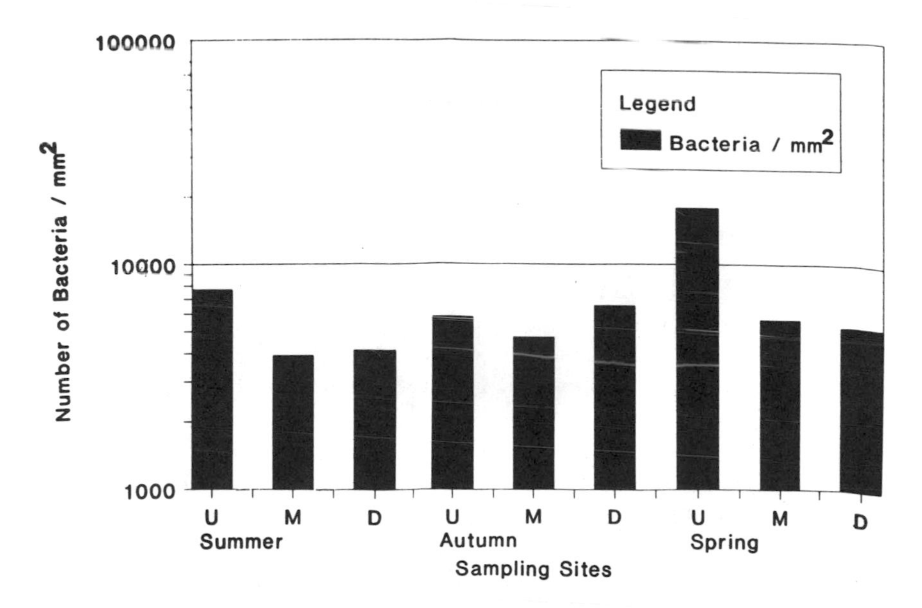

FIGURE 17. Levels of total viable bacteria per mm² surface area adsorbed on particles 10 to 30 μm in diameter in the Central School Drain. U, upstream; M, main; D, downstream.

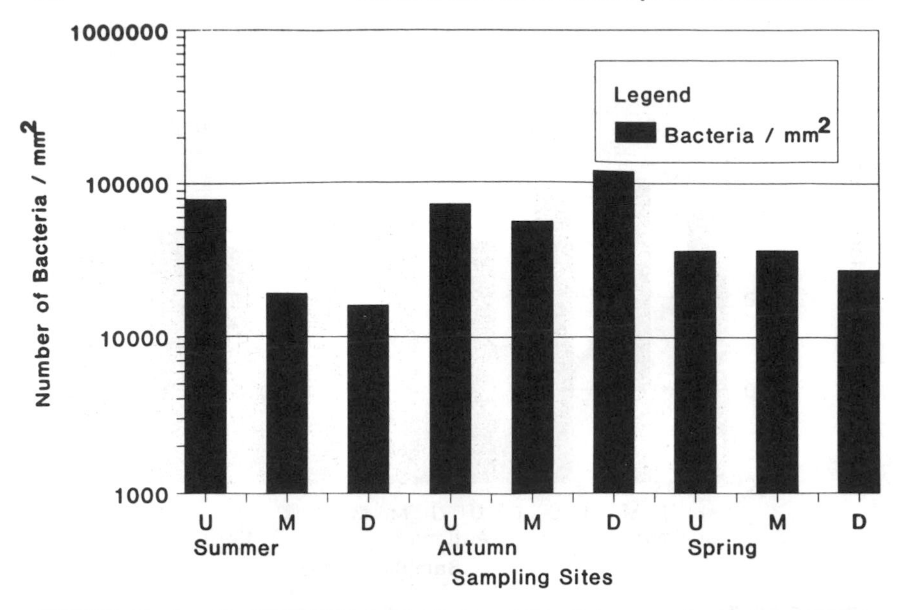

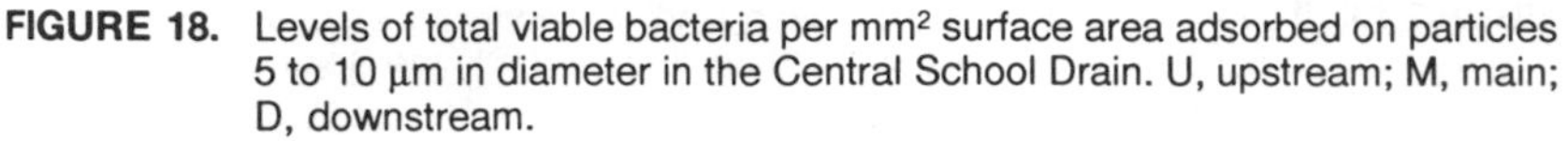
FIGURE 18. Levels of total viable bacteria per mm² surface area adsorbed on particles 5 to 10 μm in diameter in the Central School Drain. U, upstream; M, main; D, downstream.

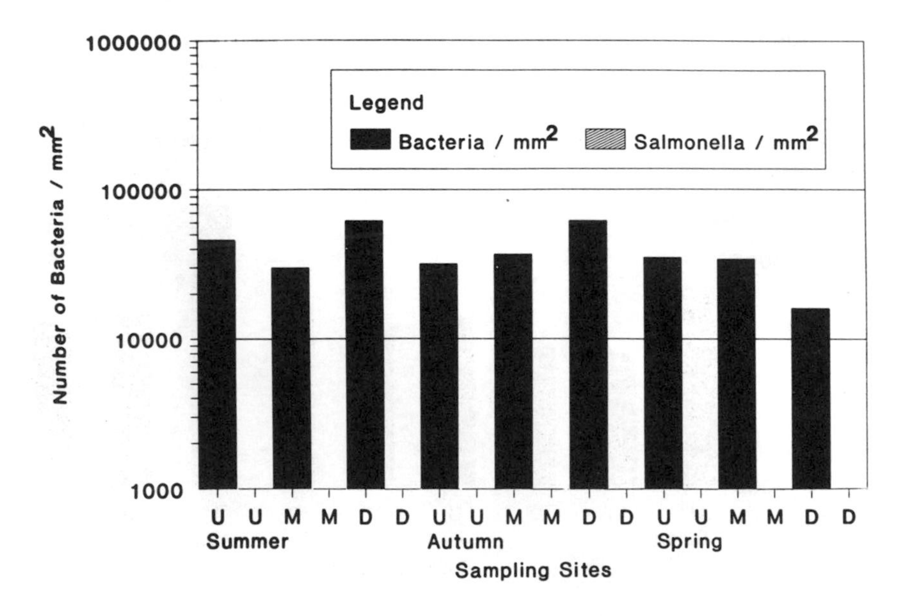

FIGURE 19. Levels of total viable bacteria and *Salmonella* per mm² surface area adsorbed on particles 1 to 5 μm in diameter in the Central School Drain. U, upstream; M, main; D, downstream.

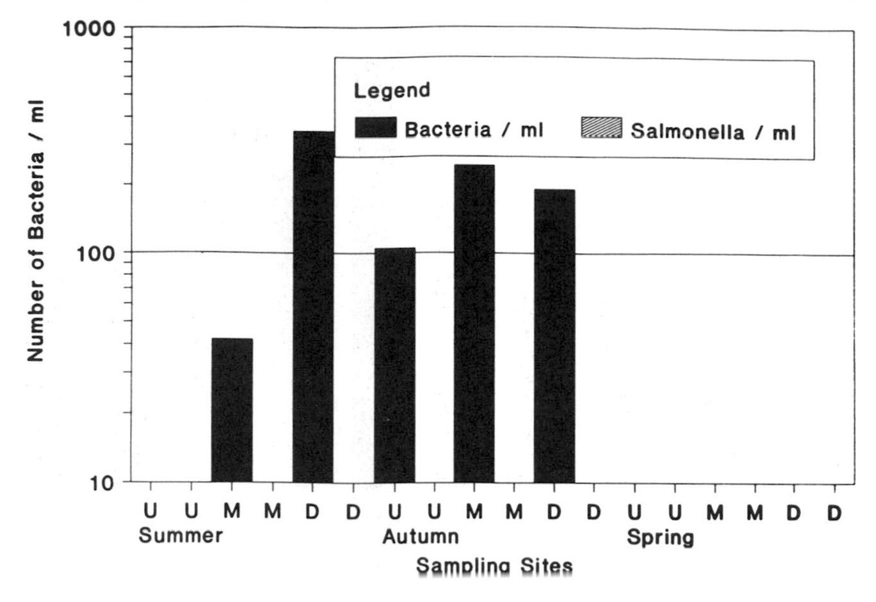

FIGURE 20. Levels of total viable bacteria and *Salmonella* that are free-floating and associated with particles 30 to 70 μm in diameter in the Central School Drain. U, upstream; M, main; D, downstream.

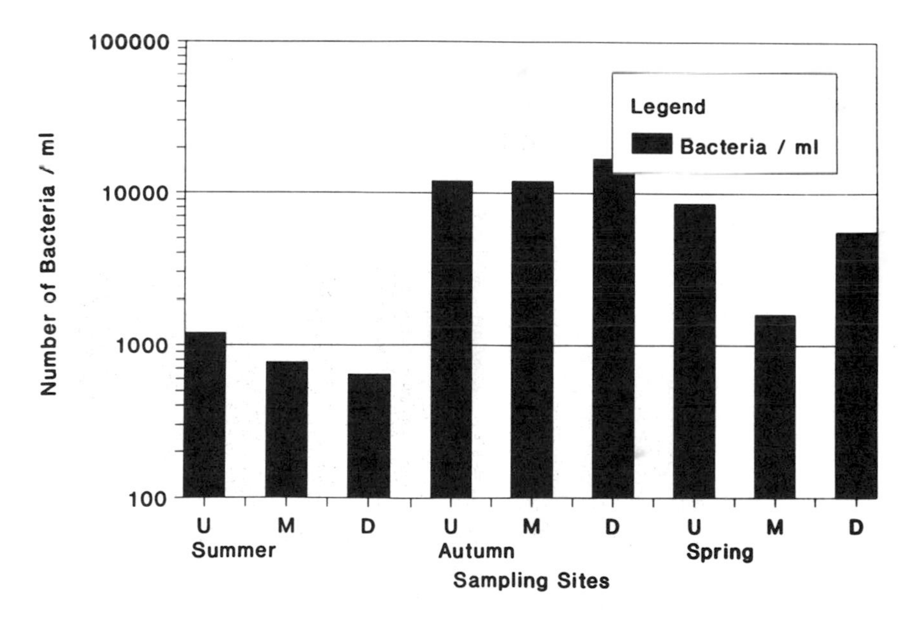

FIGURE 21. Levels of total viable bacteria that are free-floating and associated with particles 10 to 30 μm in diameter in the Central School Drain. U, upstream; M, main; D, downstream.

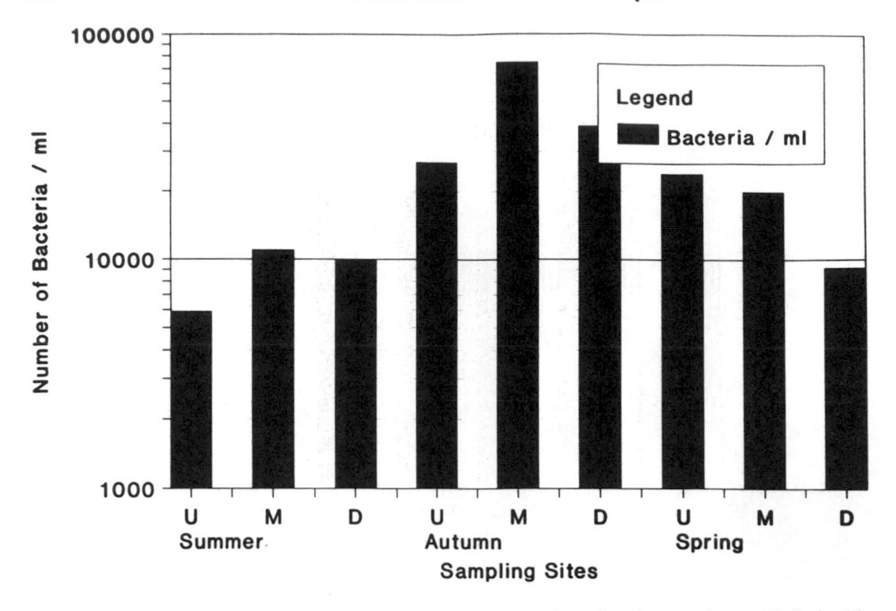

FIGURE 22. Levels of total viable bacteria that are free-floating and associated with particles 5 to 10 μm in diameter in the Central School Drain. U, upstream; M, main; D, downstream.

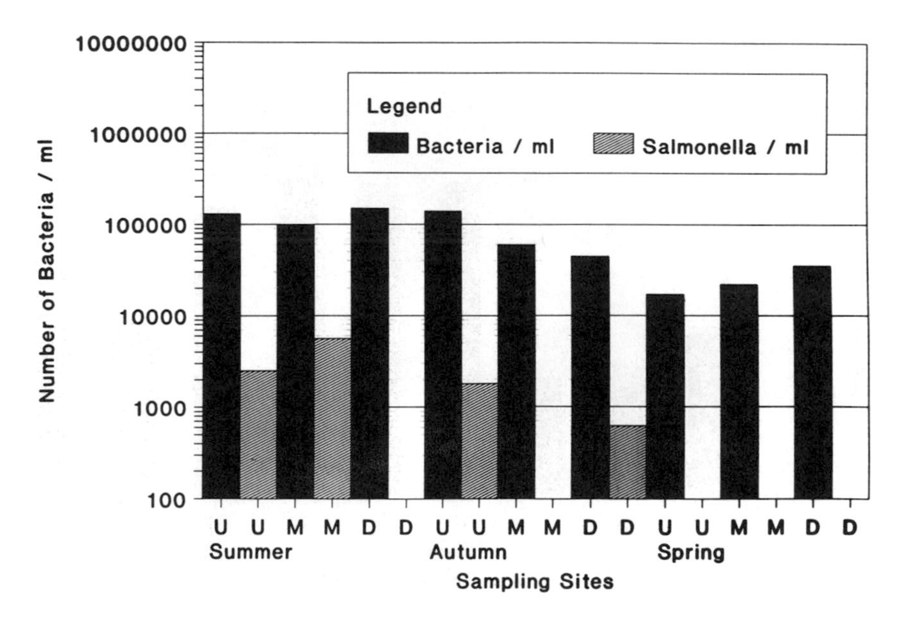

FIGURE 23. Levels of total viable bacteria and *Salmonella* that are free-floating and associated with particles 1 to 5 μm in diameter in the Central School Drain. U, upstream; M, main; D, downstream.

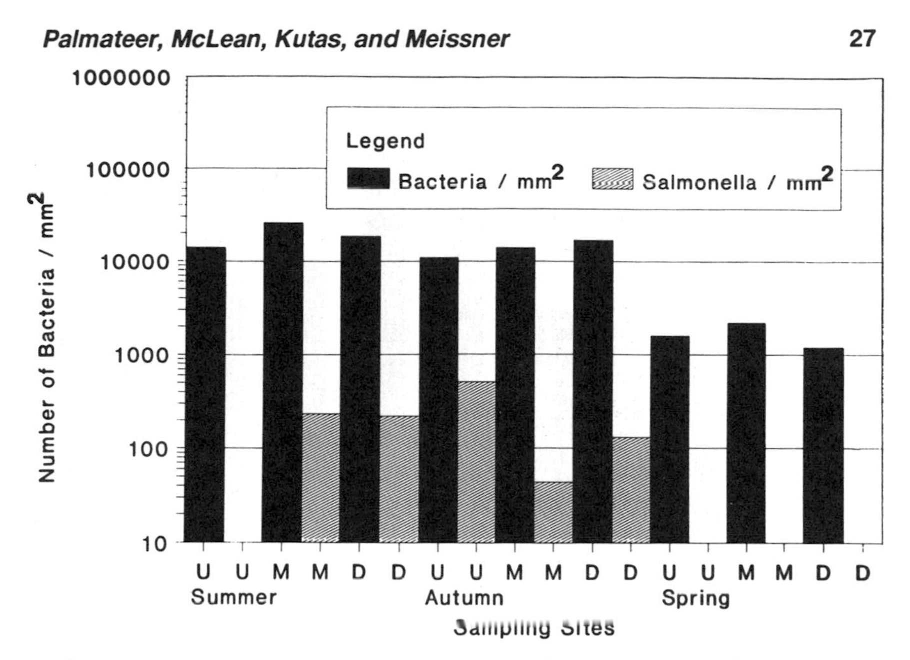

FIGURE 24. Levels of total viable bacteria and *Salmonella* per mm² surface area adsorbed on particles 30 to 70 μm in diameter in the Desjardine Drain. U, upstream; M, main; D, downstream.

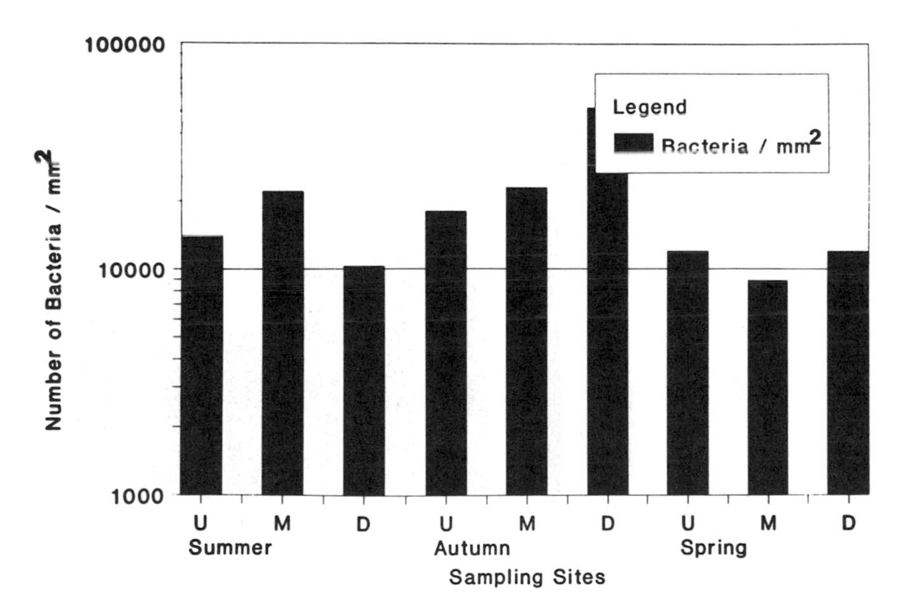

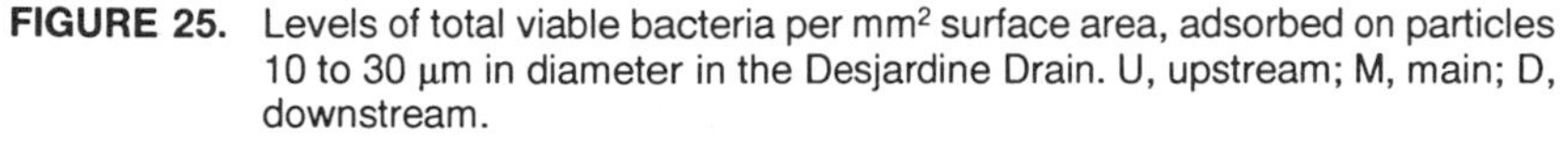

FIGURE 25. Levels of total viable bacteria per mm² surface area, adsorbed on particles 10 to 30 μm in diameter in the Desjardine Drain. U, upstream; M, main; D, downstream.

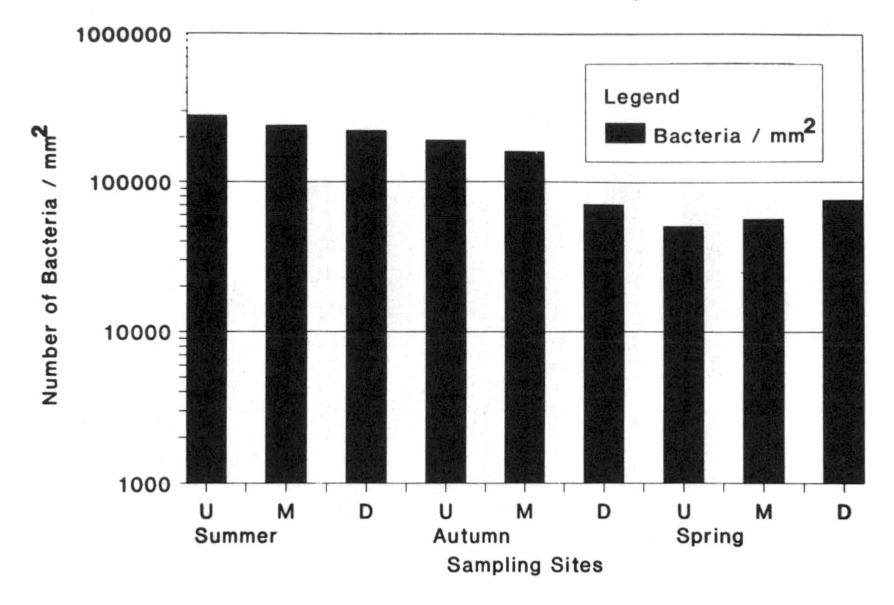

FIGURE 26. Levels of total viable bacteria per mm² surface area adsorbed on particles 5 to 10 μm in diameter in the Desjardine Drain. U, upstream; M, main; D, downstream.

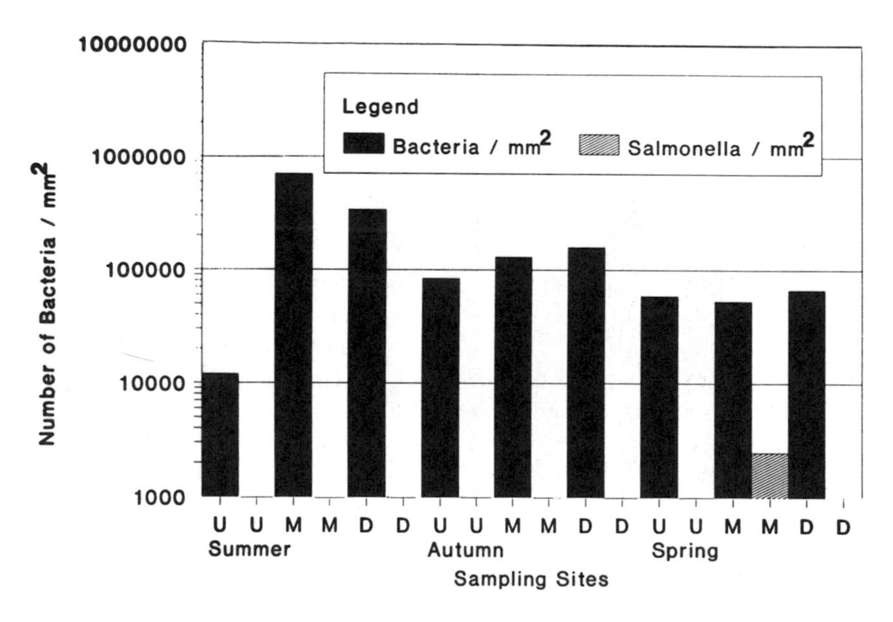

FIGURE 27. Levels of total viable bacteria and *Salmonella* per mm² surface area, adsorbed on particles 1 to 5 μm in diameter in the Desjardine Drain. U, upstream; M, main; D, downstream.

In contrast, the levels of total viable bacteria show an increase of one logarithm to 10^5 cells per mm^2 of particulate surface area. Again, a decrease in the levels of bacteria was observed in the spring, as compared to the summer and autumn levels.

Salmonella were only observed on particulates in the 1- to 5-μm diameter range, as shown in Figure 27.

Figures 28 through 31 show that the free-floating bacterial concentrations per ml vary significantly, from 1.1×10^3 to 1.8×10^5.

Free-floating *Salmonella* were associated most frequently and found at higher concentrations with the 1- to 5-μm diameter particulates than with any of the other size fractions.

The percent viability of bacterial cells sorbed to particulates or cells that were free floating and associated with various sizes of particulates was determined as an integral part of the bacterial analysis.

Table 3 shows the percent viability as the number of viable cells, divided by the total numbers of cells, multiplied by one hundred. This was calculated for both sorbed and free-floating cells. It is evident from the data that the percent viability varies with the drain being considered.

The highest percent viability, 59.2%, was observed with bacteria in the Desjardine Drain in the autumn. The average percent viability of 41.0% for the Desjardine Drain was equaled once by the average percent viability of the Arthur Vanatter Drain during the autumn. No trend was readily observed from the data other than that the average viability in the spring for each drain was the lowest while the highest average percent viability occurred in the summer for two of the three drains. The percent viability of the free-floating bacteria was generally lower than for the bacteria sorbed to particulates.

D. Bacterial Transport Study

The *Escherichia coli* NAL was recovered from the Desjardine Drain at sites shown in Figure 32. The levels in both morning and afternoon, 24 h after the discharge of the *E. coli* NAL-laden sediments into the drain, were at 10^3 cells per 100 ml. The exception occurred at the beach at Grand Bend where the cells underwent an enormous dilution when the Old Ausable River discharged into Lake Huron. The tracer bacterium remained at approximately 10^3 for the afternoon of day 2 of the study.

In Figure 33, the *E. coli* NAL concentration began to decrease at the point of insertion of the sediment — *E. coli* NAL mixture on day 3. However, the levels remained high in the drain. In addition, higher levels of tracer bacteria were recovered from the south beach station on Lake Huron.

This dispersion of the tracer bacteria continued for approximately 8 d before the levels declined in the drain. This is depicted in Figures 34 through 36 for station 2, the point of insertion and at stations 4 and 6, respectively, for a period of 85 d. The data in Table 4 substantiates the results obtained from the grab samples. The *E. coli* NAL was detectable in the Desjardine

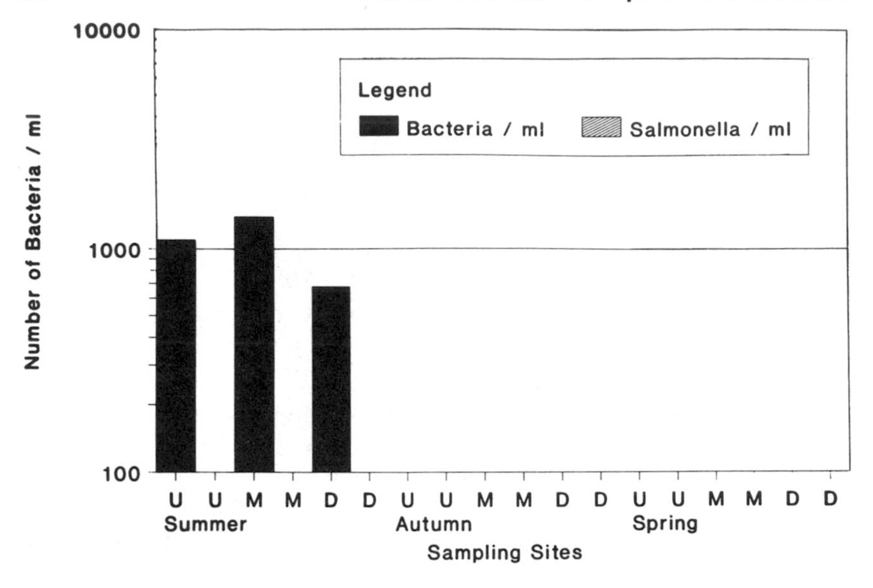

FIGURE 28. Levels of total viable bacteria and *Salmonella* that are free-floating and associated with particles 30 to 70 μm in diameter in the Desjardine Drain. U, upstream; M, main; D, downstream.

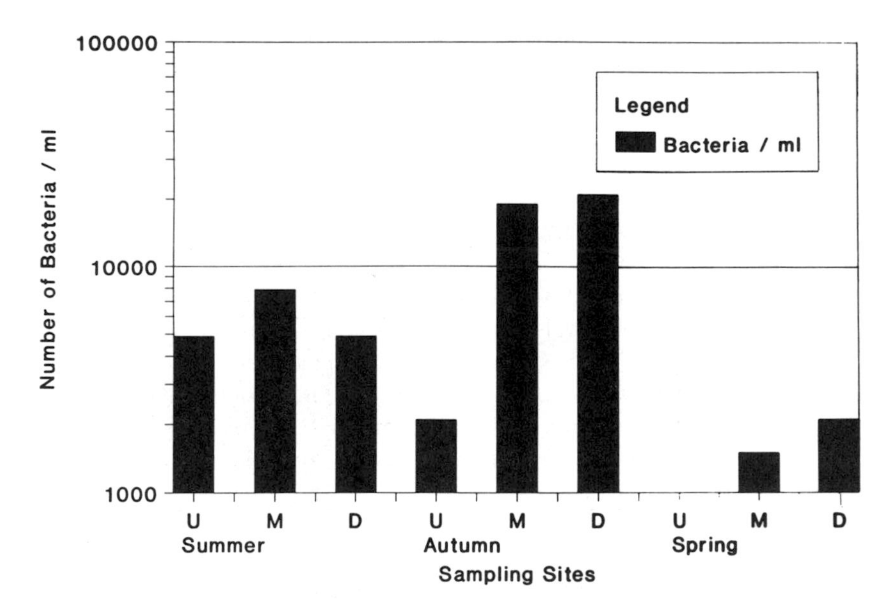

FIGURE 29. Levels of total viable bacteria that are free-floating and associated with particles 10 to 30 μm in diameter in the Desjardine Drain. U, upstream; M, main; D, downstream.

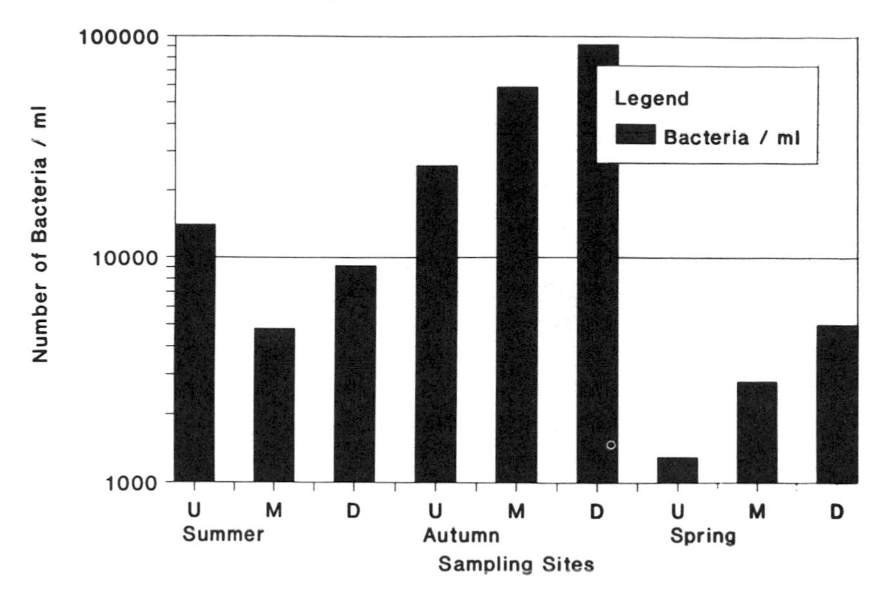

FIGURE 30. Levels of total viable bacteria that are free-floating and associated with particles 5 to 10 μm in diameter in the Desjardine Drain. U, upstream; M, main; D, downstream.

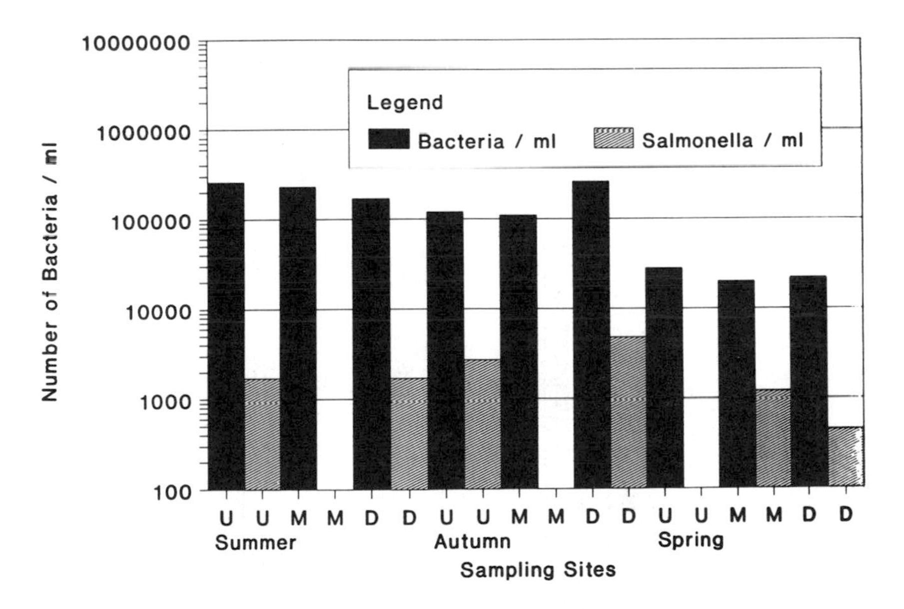

FIGURE 31. Levels of total viable bacteria and *Salmonella* that are free-floating and associated with particles 1 to 5 μm in diameter in the Desjardine Drain. U, upstream; M, main; D, downstream.

Table 3. % Viability of Total Bacteria Adsorbed to Particulates and Free-Floating at Three Study Sites in Three Seasons

Particulate size (μm)	Arthur Vanatter Drain			Central School Drain			Desjardine Drain		
	Summer	Autumn	Spring	Summer	Autumn	Spring	Summer	Autumn	Spring
Adsorbed									
30	29.8	32.5	19.1	30.0	34.5	19.2	32.2	30.7	23.2
10	35.3	38.3	20.7	34.8	25.8	23.6	33.7	32.6	24.9
5	38.2	46.7	26.6	44.5	38.9	27.0	38.9	35.4	26.8
1	27.7	45.3	46.4	35.2	35.8	19.7	59.2	32.3	35.1
Average	**32.8**	**40.7**	**28.2**	**36.1**	**33.8**	**22.4**	**41.0**	**32.8**	**27.5**
Free-floating									
30	52.4	30.9	8.25	21.0	14.9	0.0	34.2	0.0	0.0
10	26.6	43.3	15.9	32.9	33.0	21.4	35.9	56.2	11.9
5	21.5	31.5	21.3	36.0	31.3	18.7	34.5	44.3	24.0
1	30.6	41.3	17.5	23.5	39.1	15.8	81.7	31.9	20.0
Average	**32.8**	**36.8**	**15.7**	**28.4**	**29.6**	**14.0**	**34.1**	**33.1**	**14.0**

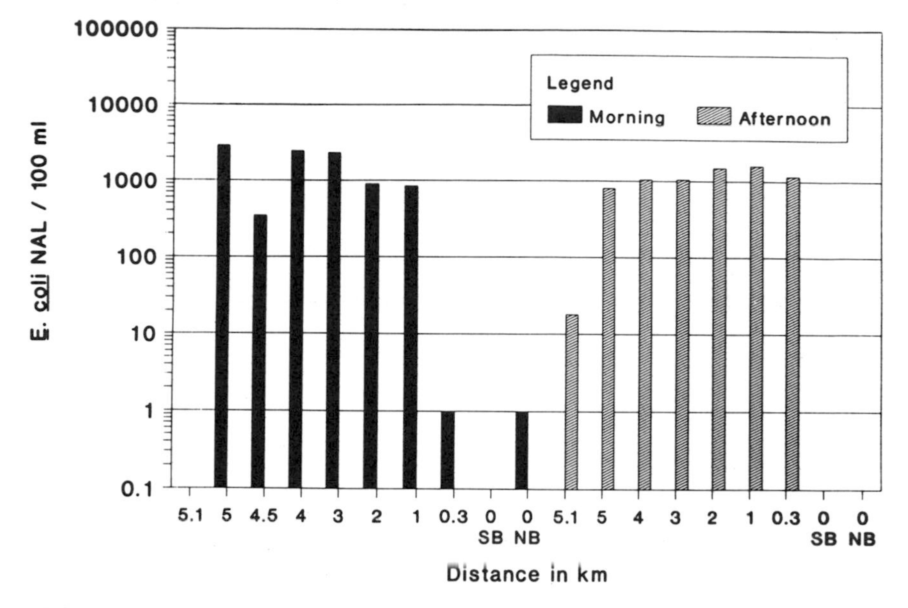

FIGURE 32. Levels of *E. coli* NAL recovered in the Desjardine Drain at ten sites on November 20, 1991. NB, north beach; SB, south beach.

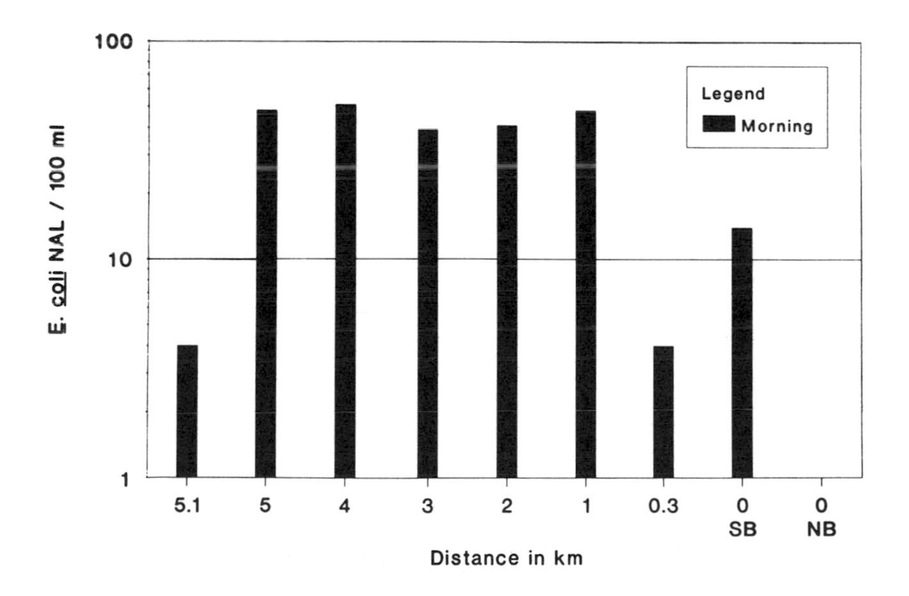

FIGURE 33. Levels of *E. coli* NAL recovered in the Desjardine Drain at 9 sites on November 21, 1991. NB, north beach; SB, south beach.

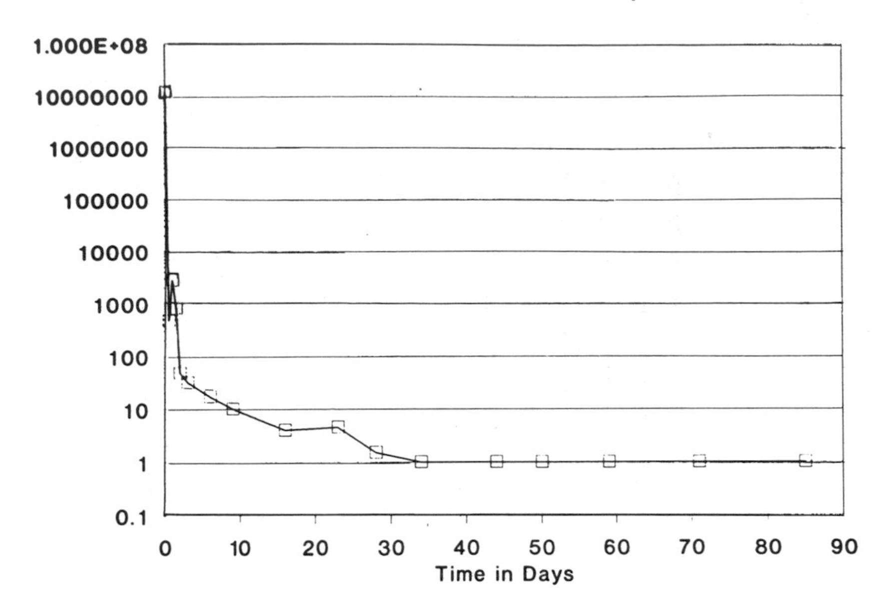

FIGURE 34. Levels of *E. coli* NAL recovered at site 2 during the entire study period of 85 d.

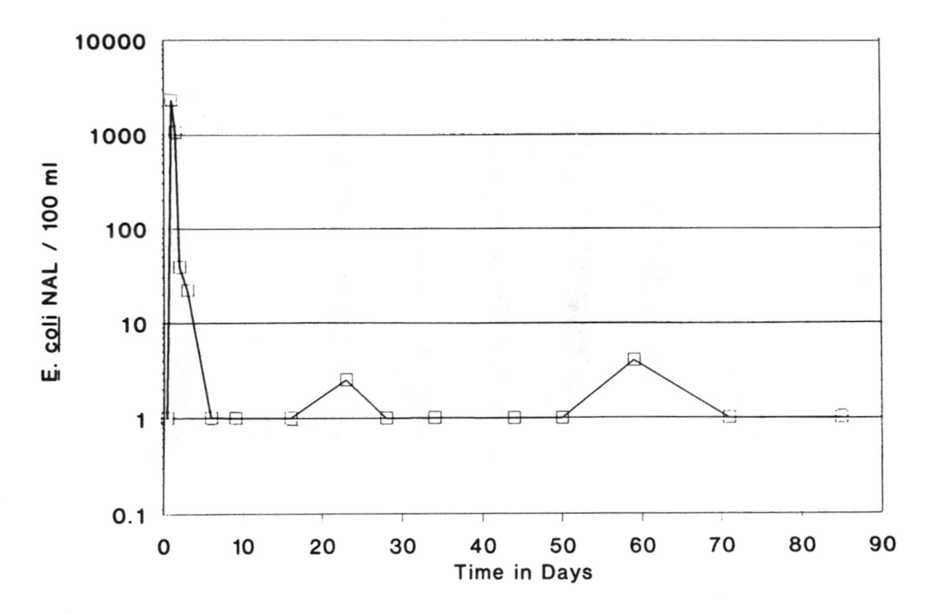

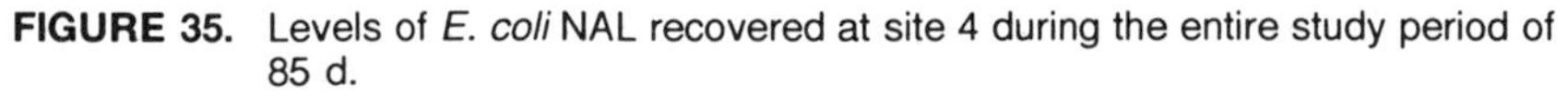

FIGURE 35. Levels of *E. coli* NAL recovered at site 4 during the entire study period of 85 d.

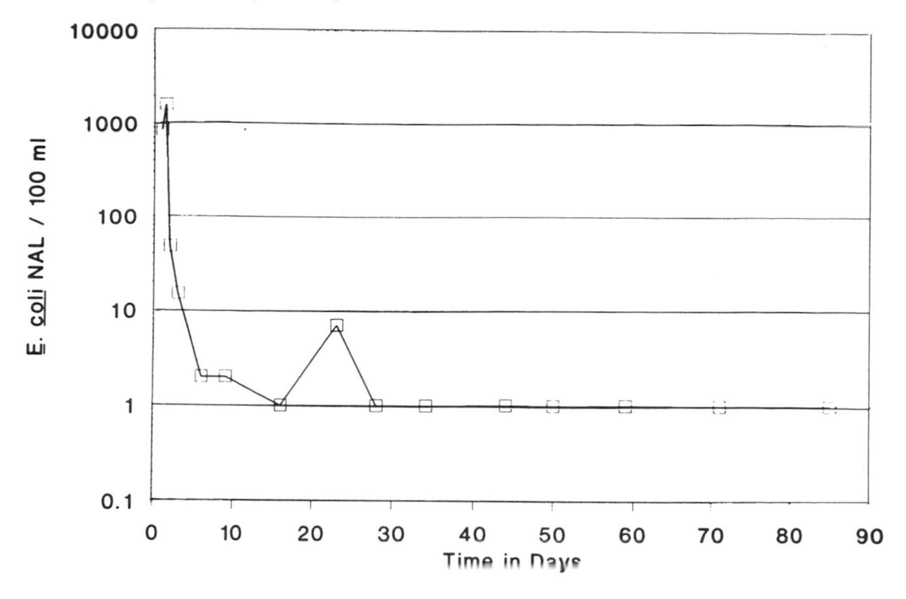

FIGURE 36. Levels of *E. coli* NAL recovered at site 6 during the entire study period of 85 d.

Drain from November 19, 1991 to January 29, 1992, using the swabs as the method of recovery.

However, sediment analyses were also conducted in the drain at the completion of the study, and sediments from sites 2, 4, and 6 contained viable *E. coli* NAL, while the upstream control station remained negative.

IV. DISCUSSION

A. Agricultural Drains

This study showed that the concentration of the pollution indicator bacteria, *E. coli*, and fecal streptococci were typical for agricultural drains in southwestern Ontario.[23]

The bacterial water quality of the Arthur Vanatter Drain during the summer was poor, with respect to the 100 *E. coli* per 100 ml bathing-beach water guideline. The *E. coli* mean of 600 bacteria per 100 ml and fecal streptococci mean of 1700 bacteria per 100 ml were both excessive. Remedial measures, such as introducing buffer strips and fencing cattle from the drain, did not reflect as benefits to the drainage water quality. A wild deer population in a wooded area bordering the northeast side of the drain may have contributed to some of the fecal pollution in the drain near the sampling sites.

The Central School Drain's bacterial water quality was the poorest of the three study drains. Cattle regularly watered in the drain, from the upstream site to the main sampling site. In addition, a tile drain discharge at the main site was known to contain milkhouse waste water and subsurface manure runoff. The mean *E. coli* and fecal streptococci values during the summer and fall were above bathing-beach bacterial water-quality guidelines.

In contrast, during the spring, the improved means for both *E. coli* and fecal streptococci in the Central School Drain reflected the fact that the cattle had been fenced from the drain, and that the milkhouse waste had been intercepted.

The Desjardine Drain bacterial water quality was slightly higher in comparison to the Central School Drain's. The mean levels of both bacterial indicators in the autumn were higher than those of the other two drains. Septage from poorly performing septic tanks and manure from cattle watering in the stream both contributed to the fecal bacterial loadings. Spring flows were high and likely diluted the concentration of fecal bacteria for the Desjardine Drain.

All three drains demonstrated significantly improved bacterial water quality in the spring.

The suspended particulates and bacterial sorption analyses, in contrast to the water-quality data, did not reflect the same variation in bacterial concentrations as did the fecal bacterial indicators. The levels of total viable and *Salmonella* bacteria sorbed to suspended particulates were consistently highest for the Desjardine Drain. Analyses of the bottom sediments for percent sand, silt, organic matter, and clay showed that the surface sediment had three times the clay content than in either of the other two drains. In addition, bacterial concentrations were highest for the 1- to 5-μm diameter size range, which is the size range nearest that of clay particulates. In reviews by both Stotzky[24] and Marshall,[9] bacterial sorption to clays, specifically montmorillonite, was demonstrated to occur to a much greater extent than to soils with a low clay content.

Burton et al.,[25] found that a variety of pathogenic bacteria, including *Salmonella* sp., survived for much longer periods when bound to clay sediments compared to their survival in the overlying water column.

Suspended drainage sediments were a combination of water-saturated soil, present in the drain from upstream erosion, and sediments native to the bottom of the drain.

The percent clay and the organic matter, both of which act as sorbents for microorganisms, were found to be much higher in the Central School Drain than in the Arthur Vanatter Drain.

The levels of both the total viable bacteria and *Salmonella* are difficult to put into perspective due to the paucity of such data, using this methodology, in the literature. Tsernoglou and Anthony,[26] using similar methods of analysis, found levels of bacteria sorbed to sediments of fresh water lakes to range from 3×10^3 to 1.5×10^4 per mm^2 of surface area, which compare with the results shown.

The total viable bacteria and *Salmonella*-detection technique employed in this study were modifications of the methods used by Kogure et al.[16] and Xu et al.,[17] and allowed differentiation of viable from nonviable bacteria. The bacterial concentrations, sorbed to particulates or free-floating, indicate that significant levels of *Salmonella* exist in these drains. In comparison to previous sanitary surveys conducted on these drains, *Salmonella* were substantially underestimated because conventional culturing techniques were used. Xu et al.[27] demonstrated the survival and viability of nonculturable *E. coli* and *Vibrio cholerae* in aquatic environments.

Clearly, fluvial sediments can transport significant loads of bacteria on particulates that are small enough (<70-μm diameter in size) to be transported during moderate stream-flow conditions.

Salomons[28] discussed various aspects of sediments and water quality and made the point that sediments that act as sorbents for various environmental pollutants serve to improve the water quality of the water column overlying the sediments. However, when these sediments become resuspended in the water column, as they often do during significant increases in the stream flow, the water quality rapidly deteriorates.

Matson et al.[29] also concluded that river sediments previously contaminated with fecal bacteria could be resuspended at some later time and cause significant deterioration in water quality of the overlying water during elevated flow periods.

B. Bacterial Transport Study

E. coli NAL were sorbed to the sediment of the Desjardine Drain that was 22% clay and had a cation exchange capacity of 25 meq. per 100 g. This sediment was reintroduced into the Desjardine Drain. Within 24 h, the bacterial-particulate mixture had reached the beaches of Lake Huron. In recent studies, Palmateer et al.[30] demonstrated that the *E. coli* NAL could travel the entire 18 km of the Desjardine Drain from its origin, which was a field tile drainage, to the discharge of the Old Ausable River into Lake Huron at the Grand Bend Beach.

In those studies, the bacterial suspension was added to the drain with no immediate involvement of sediments. Five days from the time of insertion of the tracer bacteria, both water and sediment samples of the Desjardine Drain were found to be free of any viable *E. coli* NAL. It was evident that the entire plume of tracer bacteria had passed through the Desjardine Drain to Lake Huron. If there was a residual, it was nonculturable.

In contrast, this study took 85 d for the same *E. coli* NAL, sorbed to particulates, to eventually die off and/or pass entirely from the drain.

The effect of bacteria being sorbed to nutrient-rich sediments in an agricultural drain is obvious.

Table 4. Occurence of *E. coli* NAL at Four Swab Sites in Desjardine Drain

Date	Sample Sites and Results of Analyses			
	Swab 1	Swab 2	Swab 3	Swab 4
Nov. 19. 1991	ND[1]	+	—	ND[1]
Nov. 20, 1991	—	+	+	+
Nov. 21, 1991	+	+	+	+
Nov. 22, 1991	+	+	+	+
Nov. 25, 1991	—	+	+	+
Nov. 28, 1991	—	+	+	+
Dec. 5, 1991	—	+	+	—
Dec. 12, 1991	—	+	+	+
Dec. 17, 1991	+	+	+	+
Dec. 23, 1991	—	+	—	+
Jan. 2, 1992	+	+	+	+
Jan. 8, 1992	+	+	+	+
Jan. 17, 1992	+	—	+	ND[1]
Jan. 29, 1992	—	+	+	ND[1]

[1] ND, not determined.

V. CONCLUSIONS

A. Agricultural Drains

Suspended particulates in agricultural drainage carry high levels of bacteria at 10^3 to 10^5 cells per mm^2 as a result of the sorption processes.

B. Bacterial Transport Study

The transport of the fecal-associated bacterium *E. coli*, once sorbed to suspended particulates, may travel kilometers downstream in agricultural drains to impact bathing beaches far from the point of entry of the fecal pollution in the drain.

VI. FUTURE STUDIES

Further specific study of the sorption of *E. coli*, to stream sediments, using similar procedures including monoclonal antibodies to specific pathogenic microorganisms, should be conducted to further elucidate the water quality of agricultural drains.

Knowing that sunlight can affect bacterial survival in small agricultural drains,[30] a study should be conducted to determine the deleterious impact of sunlight on bacterial survival, if tracer bacterium, such as *E. coli* NAL, is previously sorbed to sediment particulates before entering the drainage system.

REFERENCES

1. Palmateer, G. A., and Huber, D., Lake Huron beaches: factors affecting microbiological water quality in 1984, Summary Report, Ontario Ministry of the Environment, Southwest Region, Technical Support, London, 1984.
2. Palmateer, G. A. and Huber, D., Lake Huron beach study: a microbiological water quality evaluation of Grand Bend Beach and related pollution sources in 1985, Summary Report, Ontario Ministry of the Environment, Southwest Region, Technical Support, London, 1985.
3. Dean, D. M., Foran, M. E., and Fleming, R. S., Effect of manure spreading on tile drainage water quality, in *Proc. Sixth Int. Symp. Agricultural Food Processing Wastes*, Chicago, 1990, 385.
4. Bryan, F.L., Diseases transmitted by foods contaminated by wastewater, *J. Food Prot.*, 40, 45, 1977.
5. Feachem, R.G., Sanitation and disease: health aspects of excreta and wastewater management, in *World Bank Studies in Water Supply and Sanitation, No. 3*, Johns Hopkins University Press, Baltimore, 1981.
6. Dickinson, T., Sediment transport in Ontario streams, in *Managing Ontario's Streams*, FitzGibbon, J. and Mason, P., Eds., Canadian Water Resources Association, 1987, 40.
7. Gerba, C.P., Goyal, S.M., Cech, I., and Bogdan, G.F., Quantitative assessment of the adsorptive behavior of viruses to soil, *Environ. Sci. Technol.*, 15, 940, 1981.
8. Palmateer, G.A. and Hiesl, W.S., unpublished data, 1989.
9. Marshall, K.C., Clay mineralogy in relation to survival of soil bacteria, in *Annual Review of Phytopathology*, 13, Baker, K. F., Zentmyer, G. A., and Cowling, E. B., Eds., Annual Reviews Inc., Palo Alto, 1975, 357.
10. Marshall, K.C., Adsorption of microorganisms to soils and sediments, in *Adsorption of Microorganisms to Surfaces*, Bitton, G., and Marshall, K. C., Eds., John Wiley & Sons, New York, 1980, chap. 9.
11. Paerl, H.W., Microbial attachment to particles in marine and freshwater ecosystems, *Microb. Ecol.*, 2, 73, 1975.
12. Stotzky, G., Influence of clay minerals on microorganisms. III. Effect of particle size, cation exchange capacity and surface area on bacteria, *Can. J. Microbiol.*, 12, 1235, 1966.
13. Stotzky, G. and Rem, L.T., Influence of clay minerals on microorganisms. IV. Montmorillonite and kaolinite on fungi, *Can. J. Microbiol.*, 13, 1535, 1967.
14. Loeffler, A., Essex conservation rural beaches program, Summary Report, Essex Region Conservation Authority, Essex, 1990.
15. Schallenberg, M., Kalff, J., and Rasmussen, J.B., Solutions to problems in enumerating sediment bacteria by direct counts, *Appl. Environ. Microbiol.*, 55, 1214, 1989.
16. Kogure, K., Simidu, U., and Taga, N., A tentative direct microscopic method for counting living marine bacteria, *Can. J. Microbiol.*, 25, 415, 1979.
17. Xu, H.-S., Roberts, N.C., Adams, L.B., West, P.A., Siebeling, R.J., Huq, A., Huq, M.I., Rahman, R., and Colwell, R.R., An indirect fluorescent antibody staining procedure for detection of *Vibrio cholerae* serovar 01 cells in aquatic environmental samples, *J. Microbiol. Methods*, 2, 221, 1984.

18. Walker, P.H., Woodyer, K.D., and Hutka, J., Particle-size measurements by Coulter Counter of very small deposits and low suspended sediment concentrations in streams, *J. Sed. Petrol.*, 44, 673, 1974.
19. HAMES, *Handbook of Analytical Methods for Environmental Samples*, Ontario Ministry of the Environment, Rexdale, 1984.
20. Hoff, K.A., Rapid and simple method for double staining of bacteria with 4′, 6-diamidino-2-phenylindole and fluorescein isothiocyanate-labelled antibodies, *Appl. Environ. Microbiol.*, 54, 2949, 1988.
21. Johnson, G.D., Davidson, R.S., McNamee, K.C., Russell, G., Goodwin, D., and Holborow, E.J., Fading of immunofluorescence during microscopy: a study of the phenomenon and its remedy, *J. Immunol. Methods*, 55, 231, 1982.
22. Palmateer, G.A., Walsh, M.J., Kutas, W.L., and Huber, D.M., A microbiological study of recreational waters of Lake Huron at a major beach resort in Ontario, in *Abst. Annual Meeting American Society Microbiology*, Washington, D.C., 1986, 299.
23. Hocking, D.E., Rural beaches strategy program. Ausable-Bayfield Conservation Authority: Target Sub-Basin Report, Exeter, 1987.
24. Stotzky, G., Activity, ecology, and population dynamics of microorganisms in soil, *Crit. Rev. Microbiol.*, 2, 59, 1972.
25. Burton, G.A., Jr., Gunnison, D., and Lanza, G.R., Survival of pathogenic bacteria in various freshwater sediments, *Appl. Environ. Microbiol.*, 53, 633, 1987.
26. Tsernoglou, D. and Anthony, E.H., Particle size, water-stable aggregates, and bacterial populations in lake sediments, *Can. J. Microbiol.*, 17, 217, 1971.
27. Xu, H.-S., Roberts, N., Singleton, F.L., Attwell, R.W., Grimes, D.J., and Colwell, R.R., Survival and viability of nonculturable *Escherichia coli* and *Vibrio cholerae* in the estuarine and marine environment, *Microb. Ecol.*, 8, 313, 1982.
28. Salomons, W., Sediments and water quality, *Environ. Technol. Lett.*, 6, 315, 1985.
29. Matson, E.A., Hornor, S.G., and Buck, J.D., Pollution indicators and other microorganisms in river sediment, *J. Wat. Pollut. Contr. Fed.*, 50, 13, 1978.
30. Palmateer, G.A., McLean, D.E., Walsh, M.J., and Kutas, W.L., A study of contamination of suspended stream sediments with *Escherichia coli*, *Toxic Assess.*, 4, 377, 1989.

CHAPTER 2

Interactions of Bacteria with Metals in the Aquatic Environment

Philip J. Bremer and Gill G. Geesey

TABLE OF CONTENTS

0-87371-678-7/93/$0.00+$.50

I. INTRODUCTION

Historically, metals have been discharged into bodies of water on the assumption that they would either be diluted to nontoxic levels or form nonbiologically active complexes in the sediments. In the early sixties, this was dramatically and tragically proven to be a false assumption when bacteria, present in the sediments of Minamata Bay (Japan), methylated inorganic mercury, which had been released into the bay in effluents from a factory using mercuric sulphate catalysts in acetaldehyde production. The methylated mercury accumulated in fish and shellfish which were eaten by the local inhabitants. As a consequence, by 1975, 115 people had died and many were left paralyzed. This incident focused attention on the deleterious effects that can occur when heavy metals are released into the aquatic environment.[1-3]

It is now known that when metals enter bodies of water, a number of complex events occur, with the result that the metals are either deposited to the sediments or remain in the water column either in suspension or in solution. The metals may also find their way into the biota.[4] Partitioning is controlled by numerous physical, chemical, and biological characteristics of the ecosystem. These include the speciation and concentration of the metal, the composition of the effluent and natural waters with regard to organic and inorganic ligands, salinity, pH, E_h, temperature, microbial activity, and the nature of the biota.[5]

Of the various metal complexing agents in aquatic systems, microorganisms and their constituent polymers are among the most efficient scavengers of metallic ions. For this reason, it is important to develop a better understanding of the reactivity between metals and natural microbial populations.

Metals enter aquatic habitats from a number of natural and anthropogenic sources. Natural sources include metals in the earth's crust entering via: run-off following rain, land erosion, volcanic activity, or wind-blown dust.[6] The quantity of some metals entering from natural sources is now exceeded in certain areas by that entering due to man's activities. These discharges can create point sources of elevated metal concentrations in the sediments and water column around the discharge point and for varying distances away from it.[7]

II. SURVIVAL OF BACTERIA IN THE PRESENCE OF ELEVATED METAL CONCENTRATIONS

A. Occurrence of Metal-Resistant Bacteria in Aquatic Environments

Bacteria containing multiple resistance to heavy metals have been isolated from a wide variety of environments,[8-12] including heavy metal-polluted and nonpolluted freshwater environments.[13,14] That the ability of bacteria to grow in the presence of relatively high metal concentrations is found in a wide range of microbial species, including those isolated from unpolluted sites,[9] suggests that resistance to high metal concentrations is not necessarily a result of genetic change but may result from intrinsic properties of the bacterium, such as the production of copious quantities of extracellular polymeric substances (EPS).[15-17] Thus, the ability of a microorganism to survive and reproduce in a metal-contaminated environment is dependent, to varying extents, on both genetic and physiological processes.[17]

A number of studies have focused on the association of plasmids with heavy metal resistance in bacteria isolated from polluted and unpolluted environments.[8-11,14,15,18] It is considered that plasmids play an important ecological role in natural bacterial populations. Bacteria, supplied with additional and transferable plasmid-encoded properties gain selective advantage over other organisms that lack the useful plasmid-encoded traits. Transfer of plasmids to other taxa accelerates horizontal evolution by spreading useful properties within bacterial communities. Transfer may also provide the genetic prerequisites for bacterial life in extreme environments, such as in metal-polluted aquatic habitats.[10]

It has been reported that bacteria isolated from polluted environments have a higher frequency of plasmids than those isolated from similar unpolluted sites;[18] however, other studies report no differences in plasmid frequency in bacteria isolated from polluted or nonpolluted sites.[14] It is, however, recognized that metal-resistant strains isolated from environmental or clinical sources generally have the genes conferring metal resistance on plasmids or on transponsons, or at least the genetic determinants are homologous to those found on plasmids.[9] Chromosomal mutations to heavy metal resistance can be produced in the laboratory but do not generally occur in nature.[9]

Research on the properties of plasmids has, up to recently, been conducted mainly on plasmids associated with bacteria isolated from clinical environments.[9,14] Genetic information on natural bacterial assemblages from different environments is rather limited. To enable researchers to fully understand the role of transferable genetic elements in the survival of bacteria in metal-polluted environments, more information is required on (1) plasmid distribution within the environment, (2) plasmid stability in nature, (3) transfer mechanisms, and (4) the mechanisms of metal resistance imparted to the bacteria by the acquisition of a particular plasmid.

B. Mechanisms that Enable Bacteria to Survive Elevated Metal Concentrations

Metal resistance in microorganisms can occur by a variety of mechanisms that decrease either the toxicity of the metal or its accessibility to the cell. Such processes include physical sequestration, exclusion, and complexation or modification/detoxification of the metal.

1. Extracellular Metal Complexation by Bacterial Exopolysaccharides

Among the mechanisms employed by bacteria to survive in high metal concentrations is the production of large amounts of high molecular weight substances exhibiting a range of solubilities in association with metals. These substances may detoxify metals due to their complexing or chelating properties, or they may effectively form a barrier around the cell, which limits the access of the metal ions to vulnerable sites within the cell. The production of an envelope structure or capsule around the cell is considered to be a ubiquitous feature of bacteria isolated from aquatic environments. The capsule, which is usually a polysaccharide with a repeating sequence of two to six sugar subunits,[19] is anchored to the bacterial outer membrane. In some instances, the oligosaccharide side chain of lipopolysaccharide in the outer membrane is believed to contribute to the capsular structure.[20] Protein components and other products of cell metabolism excreted by bacteria and trapped in the exopolymer matrix may also contribute to the overall chemical properties of the capsule.[21,22]

Exopolymers exhibit a variety of associations with the cell surface. They may form a firmly bound capsule, which often enables the cell to establish a stable orientation with respect to its environment. Cell-bound exopolymers may extend from 0.1 to 10 μm from the cell surface into the surrounding environment, creating a buffer zone between the surface of the cell and the external environment.[23,24]

In many instances, the capsular polymers maintain a more transient association with the cell and take the form of what is commonly referred to as "slime". Under these circumstances, a portion, or indeed, the bulk of the exopolymer sloughs into the surrounding environment. The network of polymers that make up the capsule or slime forms a colloid or gel phase, depending on the nature of the surrounding environment. The water enclosed by the capsule generally contributes greater than 99% of the capsule weight. Aggregation of the exopolymers may occur under certain conditions, resulting in the formation of visible flocs. This tendency for bacterial exopolymers to form a colloid or gel phase around the cell, which often results in biofilm or floc formation, suggests that these biomolecules possess properties that are different from those of other naturally occurring polymers.[25]

That bacterial polysaccharides play a role in protecting bacteria from the toxic effects of metal ions has been suspected for a number of years.[26] Studies indicated that bacterial survival rate in the presence of metals was significantly greater for mucoid cells as opposed to nonmucoid variants.[26,27] Indirect evidence also suggests that exopolysaccharides play a role in the protection of bacteria from metal ions; for example, it has been shown that the production of capsule and slime exopolymers is enhanced in the presence of metal ions.[28-31] Furthermore, there is evidence that the presence of elevated metal concentrations in the environment influences both the chemical composition of the EPS produced by the bacteria[32,33] and its settling/floccing properties.[30,34] Adsorption of metal ions by polymers has been suggested to enhance the protective capabilities of bacterial polymers by increasing their resistance to decomposition.[35-38] The increased persistence in the environment is considered to be due to an inactivation of bacterial enzymes by the heavy metals.[39]

Although different classes of biological molecules may be associated with bacterial exopolymers, the metal-binding reactions that will be considered here are restricted to those involving the polysaccharide component. Metals are electron acceptors. The most effective donor group associated with acidic capsules and slime polysaccharides is the carboxyl residue. Lone electron pairs on the carboxyl groups interact with the charge-compensating metal ions. Weak electron donors are also present on acidic and neutral polysaccharides in the form of oxygen atoms associated with the ether bond and hydroxyl residues on the sugar subunits.[40,41] Pyruvate also contains free carboxyl groups which are free to react with positive-charged molecules such as Cu ions. Smith et al.[42] have developed high performance liquid chromatography (HPLC) methods to quantify ketal-linked pyruvates present on exopolymers, and work is proceeding in our laboratory to determine the importance of these groups in metal binding.

On the basis of Rendleman's[13] interpretation of ion interactions with polysaccharides, metal binding by uncharged polysaccharides occurs as a result of coordination between the metal cation and oxyanion and hydroxyl groups on the donor molecule. The affinity exhibited by uncharged polysaccharides generally decreases with increasing ionic radius of metals.

The general metal-molecule interaction is an acid-base reaction:

$$M^{n+} + LH \rightarrow M^{n+}\text{-}L^{-1} + H^{+}$$

where the acid is represented by H^+, the metal ion by M^{n+}, and the base by L^{-1}. The binding of Cu ions to the capsule of a freshwater sediment bacterium (FRI) was shown to result in the displacement of protons that caused a shift in the pK_a of the capsule from 4.90 to 4.05.[44] An observed decrease in the conditional stability constant for the Cu-capsule complex with decreased pH also suggests that there was competition between Cu ions and protons at the site of metal binding. Although no uronic acids were detected in the capsular material, the pK_a value suggests the presence of carboxyl groups, possibly in the form of ketal-linked pyruvate residues.[45]

The interaction between copper ions and carboxyl groups on acidic polysaccharides is believed to depend on several factors: the nature of the component sugars and their relative distribution in the chain, the magnitude of the overall electrostatic field, and the ratios of copper to polymer and copper to simple supporting electrolyte.[46]

For further information on the binding of metal ions by exopolymers we direct the reader to Geesey and Jang,[25] Geesey et al.,[47] Jang et al.,[48] and Jang et al.[49]

2. Extracellular Complexation by Compounds Other than Polysaccharides

Extracellular complexation occurs when microorganisms produce metabolic products that are excreted, and whose presence near or around the cell results in the immobilization of the metal. For example, citric acid is an efficient metal chelator, and not only has it been reported to protect the bacteria from the toxic effect of free metal ions, it also renders the citric acid resistant to microbial degradation. This suggests that not only do metals have a toxic effect on microbes, but they can also influence microbial decomposition of organic compounds.[50] These observations corroborate the results of Lasik and Gordiyenko [39], which suggests that the binding of metals by polysaccharides increases the time the polymer-metal complex survives in the environment.

Iron is an essential element. Many microorganisms release various iron-binding molecules, called siderophores, which scavenge iron from the environment.[51,52] Iron limitation can increase the extracellular production of siderophores.[53] In *Anabaena* species, these can function as strong copper complexing agents. It is, therefore, conceivable that in some circumstances, siderophore excretion may impart protection from metal toxicity.[17]

3. Extracellular Precipitation and Crystallization

Many bacteria mediate reactions or produce metabolites that result in the crystallization and precipitation of metals on microbial cell surfaces. For example, sulphate-reducing bacteria are involved in the formation of sulphide deposits which contain large amounts of metals. Sulphide formation thus leads to metal removal from solution, and this is associated with tolerance in a variety of microbes. Metal-tolerant strains of *Klebsiella aerogenes* precipitate Pb, Hg, or Cd as insoluble sulphide granules on outer surfaces of cells.[54] Reactions of metal ions, including Hg^{2+}, Cd^{2+}, Cu^{2+}, and Zn^{2+}, with H_2S produced by *Clostridium cochlearium* also resulted in the formation of insoluble metal sulfides.[55] Protection of *Desulfovibrio* species from copper and zinc was correlated with increased H_2S production and formation of mineral deposits such as covellite (CuS) and sphalerite (ZnS).[12,56]

Bacterially produced H_2S has been shown to be capable of decreasing the toxicity of chromium by reducing Cr(VI) to the less toxic Cr(III) ions.[57] It has been noted that, in some cases, sulfide-producing organisms can protect sensitive organisms from the toxic effects of metals. When *Desulfovibrio desulfuricans* was grown in mixed culture with a metal-sensitive strain of

Pseudomonas aeruginosa, the latter organism could tolerate higher concentrations of mercurials than it could in pure culture. Results indicated that the H_2S produced by the sulfate reducer protected the pseudomonad.[58]

4. Alteration of Transport Mechanisms and Intracellular Reactions

A number of metals can be accumulated intracellularly by bacteria.[17] As a relationship between metal transport into microbial cells and metal toxicity is often observed,[17] any mechanism that results in a decrease in the passage of metals into the cell can constitute a resistance mechanism; such mechanisms include decreased transport across the cell wall or the occurrence of metal efflux systems.

For example, the absence of outer membrane proteins (Omp) b and c in *E. coli* K12 and Omp b in *E. coli* 13/r conferred resistance on these cells to elevated copper concentrations by presumably preventing the entry of Cu^{2+} into the cells. Similarly, a reduction in the amount of Omp F resulted in increased resistance in *E. coli* strains.[59,60] The presence, absence, or expression level of a particular protein may also increase resistance, as is the case with an *E. coli* strain that has been shown to be resistant to arsenic due to reduced uptake mediated by a change in the cells ATPase efflux system.[9]

Once inside the cell, metal ions may be compartmentalized and/or converted to more inocuous forms. Such processes can be effective detoxification mechanisms, and microbes expressing them may be able to accumulate metals to high intracellular concentrations. Examples of this type of mechanism include the formation of polyphosphate metal-sequestering compounds or the synthesis of intracellular metal-binding proteins, such as metallothioneins, which function in detoxification as well as in the storage and regulation of intracellular metal ion concentrations. Gadd[17] has suggested that such mechanisms may be temporary and precede other means of expulsion of accumulated metals from the cells.

5. Metal Transformation

The ability of bacteria to alter the chemical state of a metal not only gives them a potentially important, but little understood, role in the biogeochemical cycling and the bioavailability of many elements, but may also constitute mechanisms of tolerance.[17,61]

Metals cannot be broken down into other products, but may, as a result of biological action, undergo changes in valence and/or interconversion between organic and inorganic compounds.

Mercury has received considerable attention because of its high toxicity and has become an excellent model chemical for the study of biotransformation of metal salts. The complete detoxification of mercury involves the reduction of the inorganic (Hg^{2+}) or organomercurial (CH_3-Hg^+[R-Hg^+]) form to the less toxic elemental mercury (Hg^0). Mercuric reductase, the enzyme encoded by the *merA* gene, is involved in the conversion of Hg^{2+} to Hg^0, and the organomercurial lyase enzyme encoded by the *merB* gene is involved in the cleavage of the H_3C-Hg bond with subsequent release of Hg^{2+}.[62] These enzymes have frequently been reported to be plasmid encoded.[63]

Other detoxifying enzymes include arsenite oxidase, which catalyzes As^{3+} to As^{6+}, and chromate reductase, which catalyzes Cr^{6+} to Cr^{3+}; both of these reactions result in an end product which is considerably less toxic to the cell, and, hence, may be considered as an effective detoxification mechanism.[17,64,65]

In summary, the ability of microorganisms to survive in sediments containing high concentrations of metals allows no simple explanation. This is in large part due to the multiplicity of interactions that can occur between microbial cells, metal ions, and the environment. It has been estimated that for a given environment, traditional microbiological culture techniques isolate less than 10% of the microbiological populations. Given this constraint, it is highly likely that many mechanisms of metal resistance occur, about which we know very little. As our understanding of the mechanisms used by bacteria to survive in metal-polluted environments increases, so will our knowledge of how we can best take advantage of bacterial reactions for the benefit of man and the environment.

C. Regulation of Metal Resistance

In order to gain an understanding of cellular regulation of metals, it is informative to study the regulation of both an essential and a toxic metal. Mercury salts, for example, are toxic to all living organisms and have no known beneficial function. Copper is an essential element involved in redox reactions and is a cofactor for a number of enzyme reactions; at high concentrations, however, it can cause cell damage through modification of the active sites of cellular enzymes and the peroxidation of membranes. Bacteria, therefore, have to utilize different approaches in regulating the entry of these two elements into the cell.

Similarities exist between the biological processes to protect against excess copper or mercury in terms of binding proteins and the transport of the metals. Both metals are taken into the cell, and either detoxified, sequestered, or exported. Differences occur in cell regulation of these metals, as copper requires a two-way control, while mercuric ion resistance is only required to deal with toxicity by responding to mercuric ions at any concentration and is, therefore, a one-way control system.[66]

Resistance to mercuric ions, while mainly inducible, can also be constitutive. In contrast, resistance to copper must be inducible, as over-expression of the resistance proteins would result in available cellular copper concentrations decreasing to levels that inhibited cell metabolism. Using an *E. coli* strain isolated from a high copper environment, Lee et al.[66] determined that the level of expression of copper resistance as a function of metal concentration was linear, whereas a threshold response was found with increasing concentrations of mercury. The linear response to increasing copper concentrations was consistent with the necessity for homeostatic control of this type of resistance, as the level of resistance must correlate with the level of environmental copper. With mercury, however, a threshold response is expected because the homeostatic level of mercury is close to zero, so over-expression of the response is not an immediate problem.[66] In the *E. coli* strain mentioned above, the copper resistance was coded on a plasmid (pRJ1004),[67] and four genes have been identified in the resistance determinant *pco*.[66]

III. INFLUENCE OF BACTERIA ON THE DISTRIBUTION OF METALS IN THE ENVIRONMENT

A. Geomicrobial Processes

Processes carried out by microbes present in aquatic environments can result in the formation, concentration, dispersion, alteration, or fractionation of metals.

Bacteria can cause localized accumulation of metals by (1) binding the metals either intracellularly or extracellularly, as covered above in Section II.B., or by (2) altering the microenvironment surrounding the cell so that the metals precipitate or form insoluble complexes near or around the cells. The net effect of these localized precipitations is the manifold local increase in the accumulation of metals around or associated with the bacterial cell.

Iron sulfides such as pyrite, iron oxides such as ochre, or manganese oxides such as vernadite and psilomelane may be generated authigenically by microbes.[68] Bacteria can reduce ferric oxide or manganese dioxide to soluble compounds. Microbes may act selectively on a mixture of inorganic compounds by promoting selective chemical change of one or a few compounds of the mixture, causing a selective concentration or dispersion — for example, in the oxidation of arsenopyrite by *Thiobacillus ferrooxidans* or in the preferential reduction of Mn(IV) over Fe(III) in ferromanganese nodules.[69] Microbes can alter rock structure and transform primary minerals into secondary minerals, as in the conversion of orthoclase to kaolinite.[69]

Microbes can affect the solubility and availability of metals by promoting either the oxidation or reduction of an element; such processes may be carried out within the bacteria enzymatically or indirectly (nonenzymatically) through interaction of the metal with metabolic products.

Enzymatically, bacteria can oxidize manganous manganese (Mn^{2+}) (*Metallogenium* spp. *Hyphomicrobium* spp.) and/or iron (*Thiobacillus ferrooxidans*, *Leptospirillum ferrooxidans*) to yield some energy for the cell. Aerobic and anaerobic reduction of Cr(VI) by bacteria (*Aeromonas dechromatica*, *Flavobacterium devorans*, *Arthrobacter* spp.) have also been demonstrated, and while it is unclear if it is an enzymatic reduction in all species, in some cases it has been shown to be involved in respiration processes. Cr(VI) reduction to Cr(III) is beneficial ecologically because Cr(III) is less toxic than Cr(VI). Furthermore, Cr(III) tends to precipitate as a hydroxo compound around neutrality, which is the pH range around which all known Cr(VI) reducers operate.[69]

Indirect oxidation or reduction results from the fact that many heterotrophic bacteria, whether aerobic, facultative, or anaerobic, form significant quantities of organic acids (e.g., oxalic, citric, and gluconic acids) and/or CO_2 from the catabolism of complex molecules. Some of the CO_2 hydrolyzes to form carbonic acid in aqueous solution. Some chemolithotrophs and photolithotrophs form significant amounts of sulfuric or nitric acids, depending on the substrate they use as their source of energy and/or reducing power. Other heterotrophs, when growing at the expense of nitrogenous carbon and energy sources such as proteins or peptides, generate ammonia, which forms NH_4OH in aqueous solution. The acids and bases that are produced affect the redox potential of the environment and/or pH, which have a strong effect on metal solubility and mobility.[3] Furthermore, the utilization of organic acids present in the environment may alter the pH to such a degree that previously unfavorable reactions can now proceed. For example, the bacterial utilization of hydroxycarboxylic acids results in a rise in pH in the surrounding environment, which favors the oxidation of Mn(II).[69]

Bacterial utilization of substrates, such as oxalate, citrate, humic acids, and tannins may make available metals that were previously insoluble. Bacterial break down of substrates has been reported to result in the release of free ferrous iron, which then autooxidizes to ferric iron.[70] The production of ferric iron from the oxidation of ferrous iron at pH values above 5 usually leads to precipitation of the iron. However, the presence of chelating agents, such as humic substances or citrate, can prevent precipitation, which results in the unchelated ferric iron hydrolyzing at higher pH values and forming compounds such as ferric hydroxide. The latter is relatively insoluble and will tend to settle out of suspension or crystalize and dehydrate, forming FeOOH, goethite ($Fe_2O_3.H_2O$), or hematite (Fe_2O_3).[69]

In summary, bacteria play an important, if little understood, role in the cycling and transformation of metals within the aquatic environment. Bacteria or reactions mediated by bacterial products have the potential to make a metal more or less available to other aquatic organisms and to alter the toxicity of the metal to the biota. It is only with a full understanding of the reactions mediated by bacteria that predictions regarding the impact of pollutants on an environment can be assessed. Such predictions have serious implications for the environment and for the well-being of mankind.

B. Accumulation of Metals by Bacteria

The accumulation of metals by bacteria can occur by either passive or active mechanisms. In active accumulation, metal transformations or microbe-metal interactions are carried out by living, metabolically active cells. In passive accumulation, metals are transformed by physical-chemical actions not necessarily requiring participation by living cells.[71]

A number of the mechanisms involved in metal accumulation have been discussed in the preceding section on the tolerance of bacteria to metals. In summary, active metal accumulation often depends on the expenditure of energy and includes such processes as: (1) precipitation, (2) intracellular accumulation and complexation, (3) oxidation and reduction, and (4) methylation and demethylation. Passive immobilization of metals occurs when (1) a solubilized metal is chelated by a substance produced and excreted by a microbial cell, or (2) a metal binds to a cell surface by physical-chemical reactions.[71] Microbial cell walls are anionic due to the presence of carboxyl, hydroxyl, phosphoryl, and other negatively charged groups. Cationic metals rapidly bind to these sites by an energy-independent reaction.[56] Cell-surface binding of metals by bacteria is discussed in Chapter 3 of this book.

In studying the kinetics of metal accumulation by microorganisms, it has become apparent that the above processes can be divided in terms of rate of accumulation into two main types. The first involves a rapid nonspecific binding of the metal to cell surfaces, slime layers, extracellular matrices, etc., whereas the second, slower process involves metabolism-dependent intracellular uptake or modification of the metal.[58] It appears as if most heavy metals can be adsorbed onto the surface of both living and dead microbial cells.

The concentration of metal accumulated by the bacterial cell is dependent on the bacterium under study, the speciation of the metal, the composition of the culture medium, and growth conditions.[72,58] Rather than examine single cases in detail, we have decided to cite a few of the very many references available on metal accumulation by bacteria. Metals reported to be accumulated by bacteria include: Cd[73], Cr[74], Cu[75], Ge[76], Mo[77], Ni[78], Pb[79,80], and U.[81]

C. Economic Implications of Bacteria and Metal Interactions

1. Biomining

Microbiological bioleaching is a process of extracting metals from sulfide ores with low metal concentrations. Currently, microbiological leaching of metals from ore is practiced in dump and underground uranium and copper-leaching operations.[82,83] The bioleaching of refractory precious metal ores in which gold and silver are finely disseminated in sulfide minerals such as pyrite and arsenopyrite is becoming of greater interest as the price of these metals increases and the availability of simple, free-milling ores decreases.[84]

In the general microbial-leaching process, low-grade ore is dumped in a large pile (the leach dump), and a dilute sulfuric acid solution (pH around 2) is percolated down through the pile. The liquid coming out of the bottom of the pile, rich in minerals, is collected and transported to a precipitation plant where the metal is reprecipitated and purified. The liquid is then pumped back to the top of the pile, and the cycle is repeated. As needed, more acid is added to maintain the low pH.

The principal reactions catalyzed by bacteria are the direct oxidation of sulfide minerals and the indirect dispersion or solubilization of metal sulfides and oxides by ferric iron and/or sulfuric acid.[85] To illustrate these processes, examples are taken from Brock et al.[86] of the oxidation of two copper minerals. Microbial leaching is especially useful for copper ores because the copper sulfate, formed during the oxidation of the copper sulfide ores, is very water soluble.

The first mechanism involves the direct oxidation of copper by bacteria, such as *Thiobacillus ferrooxidans*, so that the monovalent copper in chalcocite (Cu_2S) is oxidized to divalent copper, thus removing some of the copper in the soluble form (Cu^{2+}) and forming the mineral covellite (CuS); in this reaction there is no change in the valence of sulfide. The bacteria apparently utilize the reaction Cu^+ to Cu^{2+} as a source of energy; however, it is not possible to rule out iron as a prime oxidizing agent for reduced copper.[85]

A second mechanism, and probably the most important in most mining operations, involves an indirect oxidation of the copper ore with ferric irons that were formed by the bacterial oxidation of ferrous irons. In most ore, pyrite is present and the oxidation of this pyrite leads to the formation of ferric irons. Reaction of the copper sulfide with ferric iron results in the solubilization of the copper and the formation of ferrous iron. In the presence of oxygen, at the pH values (1.5 to 3.0) involved, *T. ferrooxidans* reoxidizes the ferrous iron back to the ferric form, so that it can oxidize more copper sulfide. Thus, the process is kept going indirectly by the action of the bacterium on iron.[87]

Next to copper, the most important bioleaching process involves uranium. In most ores, the uranium occurs as a mixture of minerals containing the uranium in either the tetravalent (an insoluble oxide) or the hexavalent form containing the soluble uranyl ion (UO_2^{2+}).[86,88] Bacterial leaching of uranium also occurs via an indirect mechanism. Bacterially generated Fe^{3+} oxidizes tetravalent uranium to hexavalent uranium. The Fe^{2+} so formed is then reoxidized to the Fe^{3+} by *T. ferrooxidans*. Most uranium contains associated pyrite, which serves as a source of Fe^{3+}. It is also possible to add ferric iron to trigger the process. Once the initial reaction has occurred, the iron can continue to cycle between the oxidized and the reduced form as the uranium is oxidized and solubilized. The soluble uranyl ion formed in the process is removed from the leach solution by means of organic solvent extraction, which does not involve bacteria. *Thiobacillus ferrooxidans* can directly oxidize reduced compounds of uranium (uranous sulfate, and UO_2) without the involvement of the Fe^{3+}/Fe^{2+} couple as the electron carrier; however, this reaction is not significant during dump or heap leaching due to the abundance of pyrrhotite and pyrite present within most ores.[86,88]

Thiobacillus ferroxidans appears to be the most dominant organism in the oxidation of mineral sulfides in most acidic environments if the temperature is below 40°C.[89] Other bacteria implicated in at least some steps in the bioleaching of sulfide minerals include: *Leptospirillum ferrooxidans,* which can grow on and degrade pyrite; *T. thiooxidans,* which can influence the leaching of minerals, e.g., zinc sulfides and cadmium sulfide, but cannot directly attack the lattice crystal structure of minerals or oxidize iron;[85] and *Sulfobacillus thermosulfidooxidans,* which has been isolated from geothermal mineral-enriched areas. *Sulfobacillus thermosulfidooxidans* will grow autotrophically on iron but requires a source of reduced sulfur.[89]

While the bacteria do not appear to have to attach to the mineral surface for leaching to occur, transmission electron microscopy studies have revealed that the bacteria will selectively attach to the surfaces that act as energy sources, such as, $CuFeS_2$ and FeS_2.[90]

2. Recovery of Metals from Waste Streams

Most of the current research on metal accumulation by microorganisms has focused on the use of microorganisms as a viable commercial alternative to costly and often ineffective physical-chemical technologies for the treatment of vast quantities of waste water containing low concentrations of soluble and particulate metals.[91,92] Applications of these bioabsorbents include: (1) removal of metals from aqueous, industrial effluents for pollution control; (2) remediation of contaminated surface waters, groundwaters, and lagoons; and (3) treatment of industrial process streams for resource recovery.[71]

The first generation of biological-based products for metal removal from waste and process streams is currently in use. The systems utilize either dead or living biomass. The advantage to utilizing a living system is that it is essentially a renewable resource; however, they have proven to be difficult to regulate and are susceptible to fluctuations in the loading of the metal and to the presence of other toxic compounds that may be present in the waste-water stream.

A second type of system that is showing great promise is the use of immobilized, dead, or nonmetabolizing cells, or cellular constituents in matrices that are granular in form and are both chemically and physically stable. Once the immobilized cells are saturated, they can be removed from the waste stream, stripped of products, and regenerated for further use.[56,71] While the use of dead biomass or derived products eliminates problems of toxicity, nutrient supply, and maintenance of optimal growth conditions, living cells can exhibit a wider variety of mechanisms for metal accumulations, including transport and intracellular and extracellular precipitation as discussed above in Section II. B.

3. Corrosion

Corrosion is an electrochemical reaction generated by point heterogeneities in potential on a metal surface. Once a surface makes contact with an electrolyte, an electromotive force (EMF) is generated between two parts of the surface, which results in the formation of a corrosion cell. One site on the metal surface acts as the anode, and its atoms ionize and become soluble. Simultaneously, electrons from the ionized metal atoms migrate to the less reactive point, which acts as the cathode. The surface seeks an equipotential state, and, as this state is reached, metallic dissolution of the anode follows and continues until the EMF decreases and corrosion stops. Outside forces, of course, can intervene, and it is these intervening forces which sustain or increase corrosion.[93]

Bacteria on the surface of a metal can influence corrosion rates in a number of ways. Bacterial attachment and growth on a metal surface is usually very irregular; the bacteria form a film or consortium comprised of different organisms, which can frequently be several to hundreds of cells thick. The development of this so-called "biofilm" results in heterogeneities in microbial types and their distribution over the surface of the metal.[94] Bacterial activities within a biofilm can contribute to corrosion in many ways: (1) by the production of organic or inorganic acids; (2) bacteria can alter the E_h or electrode potential of a point on the metal surface by the production of acids or bases, or by differential binding of metals by the bacteria or their exopolymers, leading to the formation of oxygen-differential or metal-concentration cells; (3) bacteria can depolarize surfaces by oxidizing hydrogen; (4) sulfate reducers produce H_2S which, in itself, is corrosive.

The effect of bacteria on iron corrosion has been documented in two recent reviews.[95,96] The remainder of this section will be devoted to bacterially induced corrosion of copper.

Recently, bacteria have been implicated in the pitting corrosion of copper in freshwater systems.[97,98] While the exact mechanism of this corrosion process is unclear, exopolymers secreted by biofilm bacteria have been implicated. Studies carried out using copper-coated germanium (Ge) internal reflection elements (IRE) in a Fourier transform infrared (FT-IR) spectrophotometer indicated that isolated bacterial exopolymers adsorbed to copper thin films deposited on the surface of the IREs.[99,100]

Results of several studies suggest that copper is oxidized by acidic polysaccharides. X-ray photoelectron spectroscopy (XPS) demonstrated that some of the copper deposited on IREs exposed to gum arabic and alginic acid was oxidized to Cu^{2+}, copper from thin films exposed to *Alteromonas colwelliana* exopolymer was oxidized to Cu^{1+}, while copper from thin films exposed to *Alteromonas (Pseudomonas) atlantica* exopolymer displayed little oxidation and remained as Cu^{o}.[101,102]

Auger depth profiles of copper thin films exposed to various acidic polysaccharides and exopolymers from biofilm bacteria verified FT-IR and atomic absorption spectroscopic results that copper had been removed from the surface. The overall rate of copper removal based on Auger depth profile results: alginic acid > gum arabic > *A. colwelliana* exopolymer > *A. atlantica* exopolymer was consistent with results obtained by ATR/FT-IR.

The interactions of bacteria isolated from corroded copper coupons on thin films of copper deposited onto Ge IRE were evaluated nondestructively in real time by FT-IR. The films were stable in flowing or static, sterile culture medium. When exposed to and colonized by the bacterium CCI#8, which had been isolated from a pit on a corroded copper coupon, the copper thin film corroded. Corrosion was enhanced under quiescent conditions. In conjunction with corrosion of the copper thin film, FT-IR spectra indicated that there was an increase in the concentration of polysaccharide material at the copper biofilm interface.[98]

On the basis of the data presented above, deterioration of copper surfaces colonized by microbial biofilms is likely due to interactions between copper ions in equilibrium with metallic copper and the exopolysaccharides secreted by the adherent microorganisms. Acidic polysaccharides, including those secreted by biofilm-forming microorganisms, have been shown to possess high-affinity binding sites for copper ions.[44,49] It has been proposed that the complexation of copper ions by the polysaccharides reduces the free metal ion concentration at the metal surface and promotes further ionization of metallic copper in order to establish equilibrium conditions.[103] A result of copper ion complexation by acidic polysaccharides from some biofilm bacteria is the liberation of hydrogen ions.[44] The resulting increase in acidity within the biofilm is likely to promote further dissolution of metallic copper.

Acid production by bacteria has been considered one of the possible mechanisms of microbially-enhanced corrosion of metals.[104] Dissolved low-molecular weight acids, such as acetic acid, have received the greatest attention in this regard.[105] Acidic polysaccharides, however, possess properties that make them particularly important agents of metal corrosion. The interligand distance of ionizable groups on polysaccharides, such as alginic acid and exopolysaccharide of *A. atlantica*, range from 4 to 8 Å.[106] Therefore, the concentration of acidic groups associated with exopolysaccharides immobilized at or near the metal surface is likely to be considerably greater than that achieved by diffusible low-molecular weight acids excreted by some bacteria.

Acidic exopolysaccharides appear to be one of the most common metabolic products of surface-associated bacterial populations. Their widespread existence appears to stem from their participation in the adhesion of biofilm microorganisms to surfaces.[15] The results presented above suggest that acidic exopolysaccharides may also play an important role in copper corrosion.

D. The Role of Bacteria in the Passage of Metals to Foodchains

As discussed in previous sections, microorganisms not only have the ability to survive in sediments containing high concentrations of metals, but they also have the ability to accumulate metals. This raises the question of what the effect is on higher organisms, and whether they ingest bacteria which have accumulated metals. Bacteria or their extracellular secretions form the base of many foodchains.[107-109] Evidence that metals concentrated by bacteria can be passed up the foodchain has been reported from a study of a freshwater habitat.

A bacterium of the genus *Sphaerotilus* was determined to concentrate a variety of metals. Elevated metal concentrations were found in tubificid worms after ingestion of these bacteria.[110] Furthermore, ingestion of the tubificid worms by tropical fish led to increased metal concentrations being detected in the fish tissue after 4 d.[111] Further evidence proving that bacteria can play a role in incorporating metals into foodchains has come from a number of studies which have investigated the mobilization and transformation of mercury and its entry to foodchains.[112-114]

The realization that bacterial epiphytes can be major contributors to the metal concentration of plants growing in polluted freshwater environments is further evidence that bacteria may play a role in the passage of metals through foodchains. The removal of the epiphytes from the aquatic plant *Alisma plantago-aquatica* resulted in a reduction in the plants Cr level of 25 to 50%.[115] The authors postulated that animals grazing in such epiphytes in polluted environments are consuming bacteria with elevated metal concentrations. These results were supported by a study in the marine environment of the periwinkle *Melarapha cincta* grazing on the surface of the sea lettuce *Ulva*. The *Melarapha* consumed *Leucothrix*-like organisms, and elevated metal concentrations in the bacteria resulted in elevated metal concentrations in the periwinkle.[116] Similarly, periphytic bacteria found on the carapace and gills of crabs collected from near a tannery-effluent outlet were shown to be able to concentrate Cr. It was speculated that ingestion of the crabs and their associated bacteria could contribute to the passage of Cr through the foodchain.[117]

By virtue of their physical properties, bacterial exopolymers are capable of adsorbing and concentrating many metals, and this may facilitate the entry of the metal into the foodchain. It was reported that metals became more available to the sediment-feeding clam *Macoma balthica* when the metals were adsorbed to exopolymers, compared to when the metals were free in solution or bound to glass beads.[118] In a similar study, the feeding of bacterial polysaccharides with bound Cr to two common marine organisms, a polychaete species and the mudsnail *Amphibola crenata,* resulted in an increase in their Cr concentration. For the polychaete, 61% of the total Cr was shown to be contributed from ingested Cr, and autoradiographic evidence following the use of radiolabeled ^{14}C-labeled polysaccharide indicated that the polychaetes were indeed ingesting the polysaccharide.[119]

The information presented above indicates that bacteria can play a significant role in the transfer of metals to higher tropic levels and ultimately to man. These findings reinforce the concepts presented in the introduction of this chapter, that the ultimate fate of metals in aquatic environments is controlled by numerous physical, chemical, and biological characteristics of the environment. Bacteria, due to their unique ability to adapt to and survive in metal-polluted environments, play an integral role in the fate of metals deposited into aquatic environments.

IV. SUMMARY

The further study of the interactions that occur between bacteria, metals, and the environment will give us an insight into how to control bacterial/metal interactions for the benefit of man and the environment. Examples of the way in which we can utilize microbe/metal interactions for our benefit include the mining and recovery of semiprecious metals or the bioremediation of metal-contaminated sites. Other reactions of interest include those that are detrimental to man, such as microbially induced corrosion, or the bacterial methylation of inorganic mercury and the subsequent transfer of the methylated mercury to organisms in higher tropic levels and ultimately to man.

REFERENCES

1. Duffus, J. H., *Environmental Toxicology* (Resource and Environmental Science Series), Edward Arnold, London, 1980.
2. Gerlach, S. A., *Marine Pollution: Diagnosis and Therapy*, Springer-Verlag, Berlin, 1981, 218.
3. Calmano, W., Ahlf, W., and Forstner, U., Exchange of heavy metals between sediment components and water, in *Metal Speciation in the Environment*, Broekaert, J. A. C., Gucer, S., and Adams, F., Eds., Springer-Verlag, Berlin, 1990, 503.
4. Forstner, U. and Wittmann, G. T. W., *Metal Pollution in the Aquatic Environment*, Springer-Verlag, New York, 1981.
5. Moore, J. W., Influence of water movements and other factors on distribution and transport of heavy metals in a shallow bay (Canada), *Arch. Environ. Contam. Toxicol.*, 10, 715, 1981.
6. Moore, J. W. and Ramamoorthy, S., *Heavy Metals in Natural Waters: Applied Monitoring and Impact Assessment*, Moore, J. W. and Ramamoorthy, S., Eds., Springer-Verlag, New York, 1984.
7. Duedall, I. W., Ketchum, B. H., Park, P. K., and Kester, D. R., Global inputs, characteristics, and fates of ocean-dumped industrial and sewage wastes: an overview, in *Wastes in the Ocean, Vol. 1. Industrial and Sewage Wastes in the Ocean*, Duedall, I. W., Park, P. K., Ketchum, B. H., and Kester, D. R., Eds., John Wiley & Sons, New York, 1983, 3.
8. Hermansson, M., Jones, G. W., and Kjelleberg, S., Frequency of antibiotic and heavy metal resistance, pigmentation, and plasmids in bacteria of the marine air-water interface, *Appl. Environ. Microbiol.*, 53, 2338, 1987.
9. Silver, S. and Mirsa, T. K., Plasmid-mediated heavy metal resistances, *Ann. Rev. Microbiol.*, 42, 717, 1988.
10. Schutt, C., Plasmids in the bacterial assemblage of a dystrophic lake: evidence for plasmid-encoded nickel resistance, *Microb. Ecol.*, 17, 49, 1989.
11. Diels, L. and Mergeay, M., DNA probe-mediated detection of resistant bacteria from soils highly polluted by heavy metals, *Appl. Environ. Microbiol.*, 56, 1485, 1990.
12. Trevors, J. T. and Cotter, C. M., Copper toxicity and uptake in microorganisms, *J. Ind. Microbiol.*, 6, 77, 1990.

13. Burton, N. F., Day, M. J., and Bull, A. T., Distribution of bacterial plasmids in clean and polluted sites in a south Wales river, *Appl. Environ. Microbiol.*, 44, 1026, 1982
14. Belliveau, B. H., Starodub, M. E., and Trevors, J. T., Occurrence of antibiotic and metal resistance and plasmids in *Bacillus* strains isolated from marine sediment, *Can. J. Microbiol.*, 37, 513, 1991.
15. Costerton, J. W., Geesey, G. G., and Cheng, K-J., How bacteria stick, *Sci. Am.*, 238, 86, 1978.
16. Duxbury, T. and McIntyre, R., Population density-dependent metal tolerance: one possible basis and its ecological implications, *Microb. Ecol.*, 18, 187, 1989.
17. Gadd, G. M., Metal tolerance, in *Microbiology of Extreme Environments*, Edwards, C., Ed. McGraw-Hill, New York, 1990, 178.
18. Hada, H. S. and Sizemore, R. K., Incidence of plasmids in marine *Vibrio* spp. isolated from an oil field in the northwestern Gulf of Mexico, *Appl. Environ. Microbiol.*, 50, 1262, 1985.
19. Sutherland, I. W., Biosynthesis of microbial exopolysaccharides, *Adv. Microb. Physiol.*, 23, 79, 1972.
20. Shands, J. W., Localization of somatic antigen in Gram-negative bacteria using ferritin antibody conjugates, *Ann. N.Y. Acad. Sci.*, 133, 292, 1966.
21. Povoni, J. L., Tenney, M. W., and Echelberger, W. F., Bacterial exocellular polymers and biological flocculation, *J. Water Pollut. Contr. Fed.*, 44, 414, 1972.
22. Corpe, W. A., Metal binding properties of surface materials from marine bacteria, *Dev. Ind. Microbiol.*, 16, 249, 1975.
23. Sutherland, I. W., Bacterial exopolysaccharides — their nature and production, in *Surface Carbohydrates of the Procaryotic Cell*, Sutherland, I. W., Ed., Academic Press, London, 1977, 27.
24. Cretney, W. J., MacDonald, R. W., Wong, C. S., Green, D. R., Whitehouse, B., and Geesey, G. G., Biodegradation of a chemically dispersed oil, in *Proc. 1981 Oil Spill Conf.*, Atlanta, GA, 1981, 37.
25. Geesey, G. G. and Jang, L., Interactions between metal ions and capsular polymers, in *Metal Ions and Bacteria*, Beveridge, T. J. and Doyle, R. J., Eds., John Wiley & Sons, New York, 1989, chap. 11.
26. Bitton, G. and Freihofer, V., Influence of extracellular polysaccharides on the toxicity of copper and cadmium toward *Klebsiella aerogenes*, *Microb. Ecol.*, 4,119, 1978.
27. Aislabie, J. and Loutit, M. W., Accumulation of Cr(III) by bacteria isolated from polluted sediment, *Mar. Environ. Res.*, 20, 221, 1986.
28. Wilkinson, J. F. and Stark, G. H., The synthesis of polysaccharide by washed suspensions of *Klebsiella aerogenes*, *Proc. Roy. Physical Soc.*, Edinburgh, 25, 35, 1956.
29. Corpe, W. A., Factors influencing growth and polysaccharide formation by strains of *Chromobacterium violaceum*, *J. Bacteriol.*, 88, 1433, 1964.
30. Bremer, P. J. and Loutit, M. W., The effect of Cr(III) on the form and degradability of a polysaccharide produced by a bacterium isolated from a marine sediment, *Mar. Environ. Res.*, 20, 249, 1986.
31. Beech, B., Gaylarde, C. C., Smith, J. J., and Geesey, G. G., Extracellular polysaccharides from *Desulfovibrio desulfuricans* and *Pseudomonas fluorescence* in the presence of mild and stainless steel, *Appl. Microbiol Biotechnol.*, 35, 65, 1991.

32. Appanna, V. D. and Preston, C. M., Manganese elicits the synthesis of a novel exopolysaccharide in an arctic *Rhizobium*, *FEBS Lett.*, 215, 79, 1987.
33. Annison, G. and Couperwhite, I., Consequences of the association of calcium with alginate during batch culture of *Azotobacter vinlandii*, *Appl. Microbiol. Biotechnol.*, 19, 321, 1984.
34. Ferala, N. F., Champlin, A. K., and Fekete, F. A., Morphological differences in the capsular polysaccharide of nitrogen-fixing *Azotobacter chroococcum* B-8 as a function of iron and molybdenum starvation, *FEMS Microbiol. Lett.*, 33, 137, 1986.
35. Martin, J. P., Decomposition and binding action of polysaccharides in soils, *Soil Biol. Biochem.*, 3, 33, 1972.
36. Martin, J. P., Ervin, J. O., and Richards, S. J., Decomposition and binding action in soil of mannose-containing microbial polysaccharides and their Fe, Al, Zn, and Cu complexes, *Soil Sci.*, 113, 322, 1972.
37. Hepper, C., Extracellular polysaccharides of soil bacteria, in *Soil Microbiology*, Walker, N., Ed., Halsted, New York, 1975, 93.
38. Lasik, Y. A., Gordiyenko, S. A., and Kalakhova, L., Decomposition of bacterial polysaccharides in soil, *Soil Biol.*, 151, 1979.
39. Lasik, Y. A. and Gordiyenko, S. A., Complexing of soil bacteria polysaccharides with metals, *Soviet Soil Sci.*, 9(2), 192, 1977.
40. Martell, A. E., Principles of complex formation, in *Organic Compounds in Aquatic Environments*, Faust, S. J. and Hunter, J. V., Eds., Marcel Dekker, New York, 1971, 239.
41. Brown, M. J. and Lester, J. N., Metal removal in activated sludge: the role of bacterial extracellular polymers, *Water Res.*, 13, 817, 1979.
42. Smith, J. J., Quintero, E. J., and Geesey, G. G., A sensitive chromatographic method for the detection of pyruvyl groups in microbial polymers from sediments, *Microb. Ecol.*, 19, 137, 1989.
43. Rendleman, J. A., Metal-polysaccharide complexes — Part II, *Fd. Chem.*, 3, 127, 1978.
44. Mittelman, M. W. and Geesey, G. G., Copper-binding characteristics of exopolymers from a freshwater-sediment bacterium, *Appl. Environ. Microbiol.*, 49, 846, 1985.
45. Platt, R. M., Geesey, G. G., Davis, J. D., and White, D. C., Isolation and partial chemical analysis of firmly bound exopolysaccharide from adherent cells of a freshwater sediment bacterium, *Can. J. Microbiol.*, 31, 675, 1985.
46. Manzini, G., Cesaro, A., Delben, F., Paoletti, S., and Reisenhofer, E., Copper(II) binding by natural ionic polysaccharides, Part 1, potentiometric and spectroscopic data, *Bioelectrochem. Bioenergetics* 12, 443, 1984.
47. Geesey, G. G., Jang, L., Jolley, J. G., Hankins, M. R., Iwaoka, T., and Griffiths, P. R., Binding of metal ions by extracellular polymers of biofilm bacteria, *Water Sci. Technol.*, 20, 161, 1988.
48. Jang, L. K., Brand, W., Resong, M., and Mainieri, W., Feasibility of using alginate to absorb dissolved copper from aqueous media, *Environ. Prog.*, 9, 269, 1990.
49. Jang, L. K., Harpt, N., Grasmick, D., Vuong, L. N., and Geesey, G. G., A two-phase model for determining the stability constants for interactions between copper and alginic acid, *J. Phys. Chem.*, 94, 482, 1990.

50. Brynhildsen, L. and Rosswall, T., Effects of cadmium, copper, magnesium, and zinc on the decomposition of citrate by a *Klebsiella* sp., *Appl. Environ. Microbiol.*, 55, 1375, 1989.
51. Jones, J. G., Iron transformations by freshwater bacteria, in *Advances in Microbial Ecology*, Vol. 6, Marshall, K. C., Ed., Plenum Press, New York, 1986, 149.
52. Bossier, P., Hofte, M., and Verstraete, W., Ecological significance of siderophores in soil, *Adv. Microb. Ecol.*, 10, 358, 1988.
53. Woestyne, M. V., Bruyneel, B., Mergeay, M., and Verstraete, W., The Fe^{2+} chelator proferrorosamine A is essential for the siderophore-mediated uptake of iron by Pseudomonas roseus fluorescence, *Appl. Environ. Microbiol.*, 57, 949, 1991.
54. Aiking, H., Groves, H., and Vant Riet, J., Detoxification of mercury, cadmium and lead in *Klebsiella aerogenes* NCTC 418 growing in continuous culture, *Appl. Environ. Microbiol.*, 50, 1262, 1985.
55. Pan-Hou, H. S. K. and Imura, N., Role of hydrogen sulfide in mercury resistance determined by plasmid of *Clostridium cochlearium*, *Arch. Microbiol.*, 129, 49, 1981.
56. Brierley, C. L., Brierley, J. A., and Davidson, M. S., Applied microbial processes for metals recovery and removal from wastewater, in *Metal Ions and Bacteria*, Beveridge, T. J. and Doyle, R. J., Eds., John Wiley & Sons, New York, 1989, 359.
57. Smillie, R. H., Hunter, K., and Loutit, M., Reduction of chromium(VI) by bacterially produced hydrogen sulphide in a marine environment, *Water Res.*, 15, 1351, 1981.
58. Gadd, G. M. and Griffiths, A. J., Microorganisms and heavy metal toxicity, *Microb. Ecol.*, 4, 303, 1978.
59. Rouch, D., Camakaris, J., Lee, B. T. O., and Luke, R. K. J., Inducible plasmid-mediated copper resistance in *Escherichia coli*, *J. Gen. Microbiol.*, 131, 939, 1985.
60. Goodson, M. and Rowbury, R. J., Copper sensitivity in an envelope mutant of Escherichia coli and its suppression by Col V, I-K94, *App. Microbiol.*, 3, 35, 1986.
61. Summers, A. O., Genetic adaptations involving heavy metals, in *Current Perspectives in Microbial Ecology*, Klug, M. F. and Reddy, C. A., Eds., ASM, Washington D.C., 1984, 94.
62. Tsai, Y. and Olson, B. H., Effects of Hg^{2+}, CH_3-Hg^+, and temperature on the expression of mercury resistance genes in environment bacteria, *Appl. Environ. Microbiol.*, 56, 3266, 1990.
63. Surowitz, K. G., Titus, J. A., and Pfister, R. M., Effects of cadmium accumulation on growth and respiration of a cadmium-sensitive strain of *Bacillus subtilis* and a selected cadmium resistant mutant, *Arch. Microbiol.*, 140, 107, 1984.
64. Wood, J. M. and Wang, H. K., Microbial resistance to heavy metals, *Environ. Sci. Technol.*, 17, 582, 1983.
65. Freeman, M. C., Aggett, J., and O'Brien, G., Microbial transformations of arsenic in Lake Ohakuri, New Zealand, *Water Res.*, 20, 283, 1986.
66. Lee, B. T. O., Brown, N. L., Rogers, S., Bergemann, A., Camakaris, J., and Rouch, D. A., Bacterial response to copper in the environment: copper resistance in *Escherichia coli* as a model system, in *Metal Speciation in the Environment*, Broekaert, J. A. C. and Adams, F., Eds., Springer-Verlag, Berlin, 1990, 625.

67. Tetaz, T. J. and Luke, R. K. J., Plasmid-controlled resistance to copper in *Escherichia coli*, *J. Bacteriol.*, 154, 1263, 1983.
68. Lowenstam, H. A., Minerals formed by organisms, *Science*, 211, 1126, 1981.
69. Ehrlich, H. L., Geomicrobiology, Marcel Dekker, New York, 1990, 646.
70. Kullmann, K-H and Schweisfurth, R., Eisenoxydierende, staebchenfoermige bakterien. II. Quantitative untersuchungen zum Stoffwechsel und zur eisenoxydation mit Eisen(II)-oxalat, *Z. Allg. Mikrobiol.*, 18, 321, 1978
71. Brierley, C. L., Metal immobilization using bacteria, in *Microbial Mineral Recovery*, Ehrlich, H. L. and Brierley, C. L., Eds., McGraw-Hill, New York, 1990, 303.
72. Ramamoorthy, S. and Kushner, D. J., Binding of mercuric and other heavy metal ions by microbial growth media, *Microb. Ecol.*, 2, 162, 1975.
73. Burke, B. E., Wing Tsang, K., and Pfister, R. M., Cadmium sorption by bacteria and freshwater sediment, *J. Ind. Microbiol.*, 8, 201, 1991.
74. Loutit, M., Bremer, P., and Aislabie, A., The significance of the interactions of chromium and bacteria in aquatic habitats, in *Chromium in the Natural and Human Environments*, Nriagu, J. O. and Nieboer, E., Eds., John Wiley & Sons, New York, 1988, 317.
75. Cotter, C. and Trevors, J. T., Copper adsorption by *Escherichia coli*, *Syst. Appl. Microbiol.*, 10, 313, 1988.
76. Chmielowski, J. and Klapcinska, B., Bioaccumulation of germanium by *Pseudomonas putida* in the presence of two selected substrates, *Appl. Environ. Microbiol.*, 51, 1099, 1986.
77. Stojkovski, S., Magee, R. J., and Liesegang, J., Molybdenum binding by *Pseudomonas aeruginosa*, *Aust. J. Chem.*, 39, 1205, 1986.
78. Rudd, T., Sterrit, R. M., and Lester, J. N., Mass balance of heavy metal uptake by encapsulated cultures of *Klebsiella aerogenes*, *Microb. Ecol.*, 9, 261, 1983.
79. Harvey, R. W. and Leckie, J. O., Sorption of lead onto two Gram-negative marine bacteria in seawater, *Mar. Chem.*, 15, 333, 1985.
80. Faison, B. D., Cancel, C. A., Lewis, S. N., and Adler, H. I., Binding of dissolved strontium by micrococcus luteus, *Appl. Environ. Microbiol.*, 56, 3649, 1990.
81. Marques, A. M., Bonet, R., Simon-Pujol, M. D., Fuste, M. C., and Congregado, F., Removal of uranium by an exopolysaccharide from *Pseudomonas* sp., *Appl. Microbiol. Biotechnol.*, 34, 429, 1990.
82. Tuovinen, O. H., Biological fundamentals of mineral leaching processes, in *Microbial Mineral Recovery*, Ehrlich, H. L. and Brierley, C. L., Eds., McGraw-Hill, New York, 1990, chap. 3.
83. Ahonen, L. and Tuovinen, O. H., Temperature effects on bacterial leaching of sulfide minerals in shake flask experiments, *Appl. Environ. Microbiol.*, 57, 138, 1991.
84. Lawrence, R. W., Biotreatment of gold ores, in *Microbial Mineral Recovery*, Ehrlich, H. L. and Brierley, C. L., Eds., McGraw-Hill, New York, 1990, chap. 6.
85. Lundgren, D. G., Valkova-Valchanova, M., and Reed, R., Chemical reactions important in bioleaching and bioaccumulation, in *Biotechnology For the Mining, Metal-Refining, and Fossil Fuel Processing Industries*, Ehrlich, H. L. and Holmes, D. S., Eds., John Wiley & Sons, New York, 1986, chap 7.
86. Brock, T. D., Smith, D. W., and Madigan, M. T., *Biology of Microorganisms*, 4th ed., Prentice-Hall, Englewood Cliffs, NJ, 1984.
87. Natarajan, K. A., Electrochemical aspects of bioleaching of base-metal sulphides, in *Microbial Mineral Recovery*, Ehrlich, H. L. and Brierley, C. L. Eds., McGraw-Hill, New York, 1990, chap 4.

88. McCready, R. G. L. and Gould, W. D., Bioleaching of uranium, in *Microbial Mineral Recovery*, Ehrlich, H. L. and Brierley, C. L. Eds., McGraw-Hill, New York, 1990, chap. 5.
89. Norris, P. R., Acidophilic bacteria and their activity in mineral sulfide oxidation, in *Microbial Mineral Recovery*, Ehrlich, H. L. and Brierley, C. L., Eds., McGraw-Hill, New York, 1990, chap. 1.
90. Berry, V. K. and Murr, L. E., Direct observations of bacteria and quantitative studies of their catalytic role in the leaching of low-grade, copper-bearing waste, in *Metallurgical Applications of Bacterial Leaching and Related Microbiological Phenomena*, Murr, L. E. Torma, A. E., and Brierley, V. K., Eds., Academic Press, New York, 1978, 103.
91. Norris, P. R. and Kelly, D. P., Accumulation of metals by bacteria and yeasts, in *Developments in Industrial Microbiology*, 20, Society for Industrial Microbiology, Houston, 1978, 299.
92. Brierley, J. A. and Brierley, C. L., Biological accumulation of some heavy metals — biotechnological applications, in *Biomineralization and Biological Metal Accumulation*, Westbroek, P. and de Jong, E. W., Eds., D. Reidel, Norwell, MA, 1983, 499.
93. Zajic, J. E., *Microbial Biogeochemistry*, Academic Press, New York, 1969.
94. Bremer, P. J., Geesey, G. G., and Drake, B., Atomic force microscopy examination of the topography of hydrated bacterial biofilm on a copper surface, *Curr. Microbiol.*, 24, 223, 1992.
95. Ford, T. and Mitchell, R., The ecology of microbial corrosion, in *Advances in Microbial Ecology*, 10, Marshall, K. C. Ed., Plenum Press, New York, 1990, 231.
96. Little, B. J., Wagner, P. A., Characklis, W. G., and Lee, W., Microbial corrosion, in *Biofilms*, Characklis, W. G. and Marshall, K. C., Eds., John Wiley & Sons, New York, 1990, 635.
97. Geesey, G. G. and Bremer, P. J., Applications of fourier transform infrared spectrometry to studies of copper corrosion under bacterial biofilms, *Mar. Tech. Soc. J.*, 24, 36, 1990.
98. Bremer, P. J. and Geesey, G. G., Laboratory-based model of microbiologically induced corrosion of copper, *Appl. Environ. Microbiol.*, 57, 1956, 1991.
99. Iwaoka, T., Griffiths, P. R., Kitasako, J. T., and Geesey, G. G., Copper coated cylindrical internal reflection elements for investigating interfacial phenomena, *Appl. Spectrosc.*, 40, 1062, 1986.
100. Jolley, J. G., Geesey, G. G., Hankins, M. R., Wright, R. B., and Wichlacz, P. L., *In situ*, realtime FT-IR/ATR study of the biocorrosion of copper by gum arabic, alginic acid, bacterial culture supernatant and *Pseudomonas atlantica* exopolymer, *Appl. Spectrosc.*, 43, 1062, 1989.
101. Jolley, J. G., Geesey, G. G., Hankins, M. R., Wright, R. B., and Wichlacz, P. L., Auger electron spectroscopy and X-ray photoelectron spectroscopy of the biocorrosion of copper by gum arabic, BCS and *Pseudomonas atlantica* exopolymer, *J. Surf. Interface Anal.*, 11, 371, 1988.
102. Jolley, J. G., Geesey, G. G., Hankins, M. R., Wright, R. B., and Wichlacz, P. L., Auger electron and X-ray photoelectron spectroscopic study of the biocorrosion of copper by alginic acid polysaccharide, *Appl. Surf. Sci.*, 37, 469, 1989.
103. Geesey, G. G., Mittelman, M. W., Iwaoka, T., and Griffiths, P. R., Role of bacterial exopolymers in the deterioration of metallic copper surfaces, *Mat. Perform.*, 25, 37, 1986.

104. Pope, D. H., Duquette, D. J., Johannes, A. H., and Wayner, P. C., Microbiologically influenced corrosion of industrial alloys, *Mat. Perform.*, 23, 14, 1984.
105. Little, B., Wagner, P., Gerchakov, S. M., Walch, M., and Mitchell, R., The involvement of a thermophilic bacterium in corrosion processes, *Corrosion*, 42, 533, 1986.
106. Jang. L. K., Quintero, E., Gordon, G., Rohricht, M., and Geesey, G. G., The osmotic coefficients of the sodium form of some polymers of biological origin, *Biopolymers*, 28, 1485, 1989.
107. Fenchel, T. M. and Jorgensen, B. B., in *Advances in Microbial Ecology*, Alexander, M., Ed., Plenum Press, New York, 1977, 1.
108. Hobbie, J. E. and Lee, C., Microbial production of extracellular material importance in benthic ecology, in *Marine Benthic Dynamics*, Tenore, K. R. and Coull, B. C., Eds., The Belle W. Baruch Library in Marine Science, University of South Carolina Press, Columbia, 1980, 341.
109. Sibert, J. R. and Naiman, R. J., The role of detritus and the nature of estuarine ecosystems, in *Marine Benthic Dynamics*, Tenore, K. R. and Coull, B. C., Eds., The Belle W. Baruch Library in Marine Science, University of South Carolina Press, Columbia, 1980, 311.
110. Patrick, F. M. and Loutit, M. W., Passage of metals in effluents through bacteria to higher organisms, *Water Res.*, 10, 333, 1976.
111. Patrick, F. M. and Loutit, M. W., Passage of metals to freshwater fish from their food, *Water Res.*, 12, 395, 1978.
112. Sealer, G. S., Nelson, J. D., and Colwell, R. R., Role of bacteria in bioaccumulation of mercury in the oyster *Crassostrea virginica*, *Appl. Microbiol.*, 30(1), 91, 1975.
113. Hardy, M. B. and Prabhu, N., Behavior of mercury in biosystems III, biotransferance of mercury through foodchains, *Bull. Environ. Contamin. Toxicol.*, 21, 170, 1979.
114. Berk, S. G. and Colwell, R. R., Transfer of mercury through a marine microbial foodweb, *J. Exp. Biol. Ecol.*, 52, 157, 1981.
115. Patrick, F. M. and Loutit, M. W., The uptake of heavy metals by epiphytic bacteria on *Alisma plantago-aquatica*, *Water Res.*, 11, 699, 1977.
116. Lee, Y., Patrick, F. M., and Loutit, M. W., Concentration of metals by the marine bacteria *Leucothrix* and a periwinkle *Melarapha*, *Proc. University of Otago Medical School*, New Zealand, 53, 17, 1975.
117. Johnson, I., Flower, N., and Loutit, M. W., Contribution of periphytic bacteria to the concentration of chromium in the crab *Helice crassa*, *Microb. Ecol.*, 1, 245, 1981.
118. Harvey, R. W. and Luoma, S. N., Effect of adherent bacteria and bacterial extracellular polymers upon assimilation by *Macoma balthica* of sediment-bound Cd, Zn, and Ag, *Mar. Ecol.-Progr. Ser.*, 22, 281, 1985.
119. Bremer, P. J. and Loutit, M. W., Bacterial polysaccharide as a vehicle for the entry of Cr(III) to a food chain, *Mar. Environ. Res.*, 20, 235, 1986.

CHAPTER 3

Interactions of Metal Ions with Bacterial Surfaces and the Ensuing Development of Minerals

Joel B. Thompson and Terrance J. Beveridge

TABLE OF CONTENTS

0-87371-678-7/93/$0.00+$.50

I. INTRODUCTION

Bacteria (Procaryotes) represent the oldest (3.5 to 3.8 billion years) and most abundant (10^7 cells per milliliter) life form on Earth. Since their origin during the Archean, bacteria have been quietly working at their microscopic level to transform a relatively hostile environment on the earth's surface into an environment acceptable to eucaryotic cells and eventually to higher life forms.[1] Over time, bacteria have also adapted to all known environments on earth, including extreme environments such as deep (kilometers) underground rock formations, deep-sea hydrothermal vents, hot and cold deserts, and extreme acidic and alkaline waters.[2–5] One of the main focuses of this chapter is to emphasize that bacteria are in some way or another affecting the earth's surface and its associated metal ions in all environments. For many years, we have been aware of the importance and necessity of bacteria in the global cycling of biologically important elements like oxygen, carbon, sulfur, nitrogen, and phosphorus.[6]

The knowledge of microbial gaseous oxygen generation, carbon fixation, sulfur, nitrogen, and phosphorus cycling is presently ingrained in our perception of the world. However, microbial biomineralization, or the formation of minerals by bacteria, is another important activity of microorganisms, which is just currently being recognized and documented in some detail.[6] We hope to impress on the reader that the scope of microbial biomineralization is much larger than we or anyone else initially thought; many recent studies have shown that it involves metals and minerals of tremendous range and variety, and that microbial biomineralization occurs, to some extent, in almost all natural environments on earth.

Intuitively, the antiquity (3.5 to 3.8 billion years) and ubiquity of bacteria suggest that microbial biomineralization has global consequences and has helped mold the surface of the Earth.[6] Thus, we hope to illustrate the important role that bacteria play in the biogeochemical processes occurring on and in the earth's surface.

II. BACTERIA AND THEIR SURFACES

Bacteria are truly ubiquitous organisms in nature, especially in aquatic environments like rivers, lakes, seas, and oceans. They generally exist in either a planktonic or benthic life-mode in these environments. We have found natural populations of autotrophic bacterial picoplankton with as many as 10^7 cells per milliliter in the water of some lakes. On the other hand, benthic bacterial populations commonly develop microbial biofilms on almost all submerged substrata in aquatic environments, including rocks, tree branches, bottles, and other inanimate objects. This microbial attachment and growth on submerged surfaces is very common, thereby resulting in the formation of relatively thick biofilms over time.[7-10] These natural biofilms are composed mainly of procaryotic and eucaryotic microorganisms (i.e., bacteria and algae, respectively [Figure 1]). One possible reason for microbial attachment at these fluid-solid interfaces is that the microbes take advantage of nutrient sorption at the solid surface, so they have increased metabolic activity over their planktonic counterparts.[11]

In either case, whether they are planktonic or benthic, bacteria must rely entirely on diffusion of nutrients and waste products to and from the cell for their livelihood.[6] (Unlike humans or other higher organisms, they cannot mechanically reach out and take a nutrient source, nor can they easily discard a toxic waste product.) Therefore, for efficient diffusion, it is an advantage for bacteria to be small (1 to 2 μm^3) and to have a very high surface area-to-volume ratio. This can be maximized even further by modifying their shape (e.g., from a sphere to a rod).[1] Also, as a result of their extremely small size, bacteria have a low Reynold's number of about 10^{-5}.[12,1] This implies that bacteria cannot out swim their local aqueous environment and, in fact, they must drag it around with them, no matter how fast they swim.[12,1] The drag forces are just too great to gain any microenvironment relief from swimming.

This envelope of water that bacteria drag around with them can be even considered as an extracellular component of the cell surface that also adds to the cell's overall dimensions.[1] It may be quite unique with respect to overall bulk-water chemistry and composition, and strong diffusion gradients may also exist across it. The points that we want to make are (1) bacteria are ideally designed for maximum diffusion; (2) they have the highest surface area-to-volume ratio of any other organism; and (3) a unique envelope of water exists around their periphery. Therefore, whether they like it or not, soluble metals in aqueous environments will inescapably come into contact and interact with their expansive surfaces.[1-6]

As to physical character, bacterial surfaces can come in many designs to interact with the external environment, including walls, capsules, S-layers, and sheaths. A bacterium in its natural environment will often possess a wall superimposed by a single or a multiple number of superficial layers.

FIGURE 1. SEM micrograph, a biofilm on a limestone surface from the Speed River, Guelph, Ontario. Bar = 500 nm.

A. The Various Surface Layers of Bacteria

A bacterium depends on its surface (cell wall, S-layer, capsule, or sheath) for physical and chemical protection, maintenance of cell shape, growth and division, diffusion of nutrients and wastes, adhesion to surfaces, and interactions with the environment.[1,6,13] Each bacterial surface layer represents a dynamic system whereby the different components are almost continually synthesized, assembled, modified, broken down, or sloughed off into the external milieu.[1,6,13]

1. Cell Walls

The reclassification of procaryotes into major kingdoms has been ongoing over the last decade. Originally, two so-called urkingdoms were established: the archaeobacteria and the eubacteria,[14] and, now, the archeae and the bacteria.[15] However, roughly a century ago, in 1884, Christian Gram developed a staining regimen that laid the foundation for the classification of bacteria into two major groups based on the format of their surfaces; these are the Gram-positive and Gram-negative bacteria.[1,15] Today, we still use Gram's reaction to separate eubacteria based on their response to the Gram-staining regimen. Those that retain the large crystal violet-iodine complex are categorized as Gram-positive eubacteria, whereas those that are decolorized by a secondary alcohol treatment and can be counterstained red by safranin are classified as Gram-negative eubacteria.[17,1,6] What Gram did not realize at the time was that his staining regimen actually distinguished between eubacterial cell-wall types, and that this was based on the differences in their chemical and structural organization.[18–20]

However, the Gram reaction can be unreliable when applied to certain eubacteria, and accordingly, these organisms are referred to as the Gram-variable group. These Gram-variable bacteria can stain either negatively or positively depending on growth phase or nutritional status at the time of the staining regimen.[16]

a. Gram-Positive Cell Walls *Bacillus* species cell walls have been studied extensively as a phyicochemical model system for Gram-positive walls.[21–26] *B. subtilis* walls consist of a two-component system when they are grown in the presence of phosphate and those of *B. licheniformis* are a three-component system.[27] When phosphate is available, *B. subtilis* walls consist predominately of peptidoglycan and teichoic acid. However, under phosphate limitation, teichoic acid production is repressed, and teichuronic acid is produced.[28–31] *B. licheniformis* walls consist of peptidoglycan, teichoic acid, and teichuronic acid, which altogether form the fabric of the surface.[21]

Peptidoglycan is the major wall polymer and forms the murein sacculus, which is approximately 25 peptidoglycan molecules thick.[17,27] It is also peptidoglycan that gives the bacterium its form, since it is knit into a predetermined

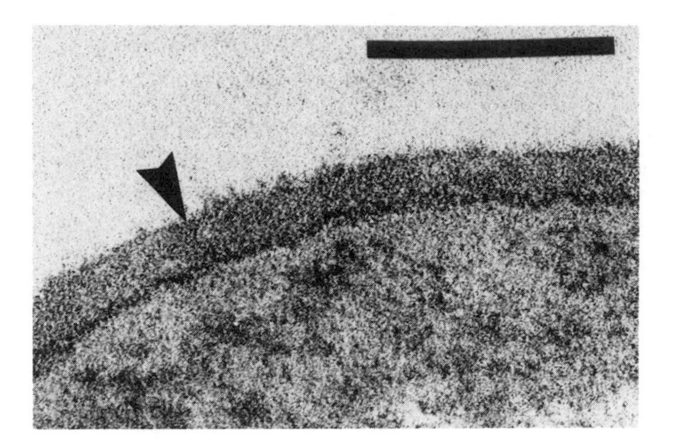

FIGURE 2. Thin section of the cell wall (arrowhead) of *B. subtilis,* which represents a Gram-positive bacterium. Bar = 100 nm.

shape and resists the strong turgor pressure exerted by the cell protoplast.[1,32,33] Much of the wall's strength and resilience comes from the interstrand cross-linkage of usually 20 to 30% of the muramyl residues present in the peptidoglycan framework. This degree of cross-linkage between the peptidoglycan polymers essentially forms a huge macromolecule through covalent bonding which entirely surrounds the entire protoplast. Commonly, peptidoglycan can account for as much as 30 to 50% of the dry weight of the Gram-positive cell wall, and this amount can vary depending on the particular growth phase and *Bacillus* species analyzed.[17] Chemically bonded to the peptidoglycan is the secondary polymer, teichoic acid. These teichoic acid chains are longer and more flexible than are the peptidoglycan chains,[27] and they can account for 50 to 60% of the wall's mass.[23,24] Teichuronic acids occupy a similar structural format, but their phosphate substituents are replaced by uronic acids.[17,31] Overall, in electron microscopic thin sections, the Gram-positive wall is an amorphous layer approximately 20 to 30 nm thick (Figure 2).

The peptidoglycan, teichoic acid, and teichuronic acid polymers within the Gram-positive wall are all capable of binding metals since they are negatively charged. The carboxyl groups present in peptidoglycan and the phosphate groups present in teichoic acid give the cell wall its anionic character.[6] When teichuronic acid is present, the uronic acid moieties contribute to the anionic character of the wall.[21,35] However, the carboxyl groups of the tetrapeptide stems of peptidoglycan dominate the net negative-charge density of the wall.[22,25]

Beveridge and Murray[22,23] unequivocally demonstrated through a number of experiments that the carboxyl groups present in the peptidoglycan are the primary sites for metal cations at circumneutral pH. In fact, for *B. subtilis* walls, their experiments suggested that most of the metal binding was due to the carboxyl groups of glutamic acid on the peptidoglycan peptide stems.[27] We must keep in mind, however, that the availability of all these wall anionic

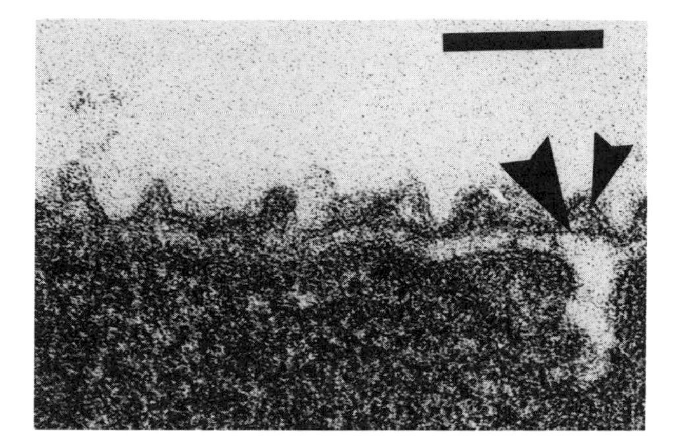

FIGURE 3. Thin section of the cell envelope of *Aquaspirillum serpens*, which represents a Gram-negative bacterium. The small arrowhead points to the outer membrane and the large arrowhead to the peptidoglycan layer. Bar = 100 nm.

groups for metals is determined by several complicating factors. These include the growth state of the cell; the degree and nature of peptidoglycan cross-linking; and the number, type, conformation, and location of additional polymers. In addition, the presence of internal ionic bonds between cationic and anionic sites within the wall fabric are also important.[13]

b. Gram-Negative Cell Walls Gram-negative cell walls are chemically and structurally different from their Gram-positive eubacterial counterparts. The most obvious difference is the lack of a thick peptidoglycan layer and the presence of a unique outer membrane within the cell envelope.[17,27] Although Gram-negative walls do contain some peptidoglycan, it only occurs as a thin single layer of one to three molecular sheets in thickness,[17,27] which is very thin compared to the previously described *B. subtilis* wall, with about 25 sheets of peptidoglycan. Therefore, Gram-negative cell walls are comprised of an outer membrane bilayer plus a thin peptidoglycan layer (Figure 3). Between the outer membrane and the plasma (cytoplasmic) membrane resides the periplasm,[17] which is best preserved by a new cryoelectron microscopical technique called freeze-substitution.[36–38] The significant reduction of peptidoglycan in Gram-negative walls means that the overall quantity of metal bound by peptidoglycan is considerably less in comparison to Gram-positive walls.[39,40]

The outer membrane is a lipid-protein bilayer with a hydrophobic core. However, it is significantly different from the plasma membrane.[41] As for the lipoid constituents, the outer membrane also contains fewer phospholipids, with more saturated fatty acids than does the plasma membrane.[42] The outer membrane contains a unique eubacterial lipid, the lipopolysaccharide (LPS), in its upper leaflet and typical phospholipids in its lower leaflet.[41,43] These unique Gram-negative LPS molecules consist of three different chemical regions: lipid

Table 1. Metal Binding by Bacterial Walls

Metal	Bacillus subtilis Native Wall[22,23]	Bacillus licheniformis Native Wall[21]	Escherichia coli Outer Membrane[113]
Na	2.697	0.910	0.200
K	1.944	0.560	0.025
Mg	8.226	0.400	0.084
Ca	0.399	0.590	0.185
Mn	0.801	0.662	0.355
Fe	3.581	0.760	0.541
Ni	0.107	0.520	0.019
Cu	2.990	0.490	ND
Au	0.363	0.031	ND

Note: *B. subtilis* and *B. licheniformis* are Gram-positive organisms; whereas *E. coli* is Gram-negative. Micromoles of metal bound per milligram dry weight of walls. ND, not determined.

A, core oligosaccharide, and O-side chain. Lipid A is hydrophobic and is, therefore, confined to the hydrophobic domain of the outer membrane. The core oligosaccharide is attached to lipid A and extends above it. Finally, an O-specific polysaccharide chain makes up the outermost substituent, is attached to the core polysaccharide, and can extend 20 to 40 nm away from the cell in *Pseudomonas aeruginosa* PA01.[44]

Hoyle and Beveridge[39,40] and Ferris and Beveridge[45,46] showed that the outer membrane is capable of binding metals, although it did not react as strongly with metal ions as did Gram-positive walls (Table 1). The primary sites for metal interaction within the outer membrane are the phosphate groups within the LPS and phospholipids.[45,46] This was conclusively demonstrated by using ^{31}P-nuclear magnetic resonance (NMR) and the paramagnetic metal ions europium and manganese on isolated outer membrane and LPS liposomes from *Escherichia coli* K12. They also showed that, although there are three carboxyl groups in the ketodeoxyoctonates of this molecule, only one is free to interact with metals; the other two are neutralized by the close proximity of amino substituents.[46]

c. Archaeobacterial Cell Walls Archaeobacterial cell walls are diverse when compared to their eubacterial counterparts. They are also chemically and structurally different from Gram-positive or Gram-negative eubacteria walls. This great variety in cell-wall structure and chemical composition among archaeobacteria probably stems from the diversity of extreme environments they seem to occupy. For example, in very hot environments, a surface capable of withstanding very high temperatures (i.e., that of *Thermoproteus* spp.) must be very different from one that resides at more ambient temperatures (i.e., *B. subtilis*).[6]

Currently, only four different wall types are known in detail for archaeobacteria.[47] They include those made of: (1) pseudomurein (e.g., Order Methanobacteriales), a heteropolymer which resembles peptidoglycan, but contains *N*-acetyltalosaminuronic acid instead of *N*-acetylmuramic acid and which lacks the D-amino acids found in eubacterial cell walls;

(2) methanochondroitin (e.g., *Methanosarcina* and *Halococcus)*, a sulfated heteropolysaccharide molecule similar to that found in animal connective tissue; (3) S-layer protein (or glycoprotein) only (e.g., *Halobacterium*); (4) no external wall layers but simply just a plasma membrane (e.g., *Thermoplasma*).[47–49]

Unfortunately, very little is presently known about the interactions between archaeobacterial walls and metal ions. However, one might expect that any free carboxyl groups present in the peptide stems of pseudomurein would contribute to their anionic character and bind metals. Additionally, the sulfated heteropolysaccharides of *Halococcus* or *Methanosarcina* should also have an anionic character, due to both carboxyl and sulfate groups. Lastly, the protein subunits of the S-layer in *Halobacterium* are known to contain covalently bonded sulfate groups that are highly anionic.[49]

2. Additional Surface Layers

In addition to a cell wall, some bacteria form other protective superficial layers, such as S-layers, capsules, and sheaths, above the cell wall. They commonly occur singly or in combination with one another and can be visualized by electron microscopy, and sometimes (e.g., capsules and sheaths), by light microscopy using differential staining techniques.[17] The common functions of these layers are protection, attachment, and uptake of nutrients and metals.

a. S-Layers Bacteria isolated from natural environments sometimes have walls that possess a layer of regularly arrayed proteins or glycoproteins on their outer surface (S-layers, Figure 4).[17,50–53] Some S-layers can consist of up to 10% of the total cellular protein produced. S-layers are found associated with all bacterial wall types, i.e., those of Gram-positive, Gram-negative, and archaeobacteria. In fact, for some archaeobacteria (e.g., *Sulfolobus* and *Methanococcus*), the S-layer can be the sole wall component.[6]

The proteins or glycoproteins, which make-up a surface array, are synthesized within the cell and transported to the surface where they self-assemble into the paracrystalline layer. Variable lattice formats have been recognized for these paracrystalline surface layers; hexagonal (p6) and tetragonal (p4) symmetries are most common; however, a few p2 examples and at least one p1 example have been documented.[52] Noncovalent interactions, such as hydrogen bonding, electrostatic interaction, and salt-bridging, are commonly invoked to provide firm attachment between neighboring subunits and to the underlying wall component.[50,51] In fact, S-layers commonly require divalent metal cations for their correct assembly on and adherence to the underlying wall component.[17,6,50–53] Calcium and magnesium cations are required most often, but sometimes Sr^{2+} can be used as a replacement ion for Ca^{2+}.[54–56] *Sporosarcina ureae*'s array requires Mg^{2+},[57] but preliminary evidence suggests that it may be responsible for most of the wall's metal-binding ability. *S. ureae* strains that can survive and grow in the presence of toxic heavy metal seem to do so by

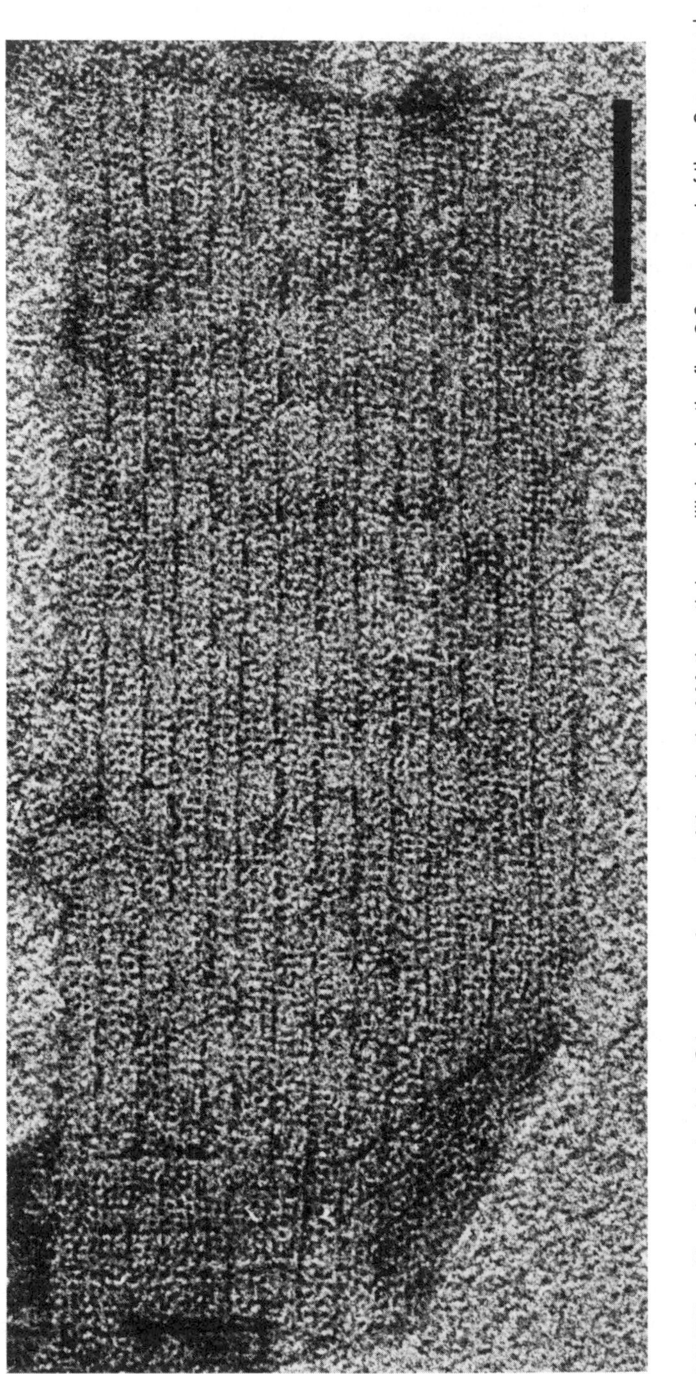

FIGURE 4. Negative stain of the S-layer on a fragment of the sheath of *Methanothrix concilii* showing the fine 2.8 nm repeat of the p2-arranged subunits. Bar = 50 nm.

using their surface array to bind and immobilize the toxic metals.[58] Once the S-layer has bound a significant amount of toxic heavy metal, it is sloughed off the cell surface and is simultaneously replaced by a new surface layer.

Synechococcus GL24, a small unicellular cyanobacterium, has been implicated in calcite mineralization in alkaline marl lakes.[59,60] Moreover, this *Synechococcus* possesses a S-layer as its outermost boundary layer that seems to act as a template which chemically binds and spatially orders Ca^{2+} so that calcite mineralization is promoted on the cell surface.[61] The S-layer lattice has p6 symmetry, and the mineralization is instigated by Ca^{2+}-binding to polar residues of the protein which surround the "empty" regions or holes in the array.[62] It is possible that the S-layer protects the cell from epicellular calcification, since old S-layer is shed into the external milieu when it becomes heavily calcified and is replaced by new S-layer.[59,61,62] Unfortunately, besides the *S. ureae* and *Synechococcus* GL24 systems, to date, little work has been done to study the influence of S-layers on cell-surface metal binding.

b. Capsules Capsules and slime layers can be distinguished from one another by the relative integrity of the exopolymers and their association with cell surfaces. Exopolymers that are tightly linked with the cell surface and that can exclude particulate negative stains such as India ink (which is used for light microscopy) are called capsules,[17] whereas, exopolymers that have a more transient association with the cell surface and that often slough off into the surrounding menstruum are called slime layers.[17,63] Capsules and slime layers are sometimes broadly grouped together and are called glycocalyes.[64] In contrast to the more rigid cell wall, bacterial capsules are highly hydrated (>90% water). They are composed of amorphous, loosely arranged exopolymers that are chemically linked to the cell surface and consist of neutral polysaccharides, charged polysaccharides, or charged polypeptides which can extend 0.1 to 10 μm from the cell wall (Figure 5)[65] They are thought to be thixotropic and to alternate between a gel to liquid state.

Capsules serve as a physical buffer between the environment and the cell. Capsular exopolymers are usually acidic in nature, since they frequently bind cationic electron microscopical dyes, such as ruthenium red, and may contain carboxyl, phosphate, and sulfate groups.[63,66–68] Capsules can contain somewhere between 5 to 25% uronic acid, and these acidic functional groups (COO^-) are capable of binding significant amounts of metal.[63] However, the anionic groups in capsules differ from those in cell walls in that they are not rigidly locked collectively by covalent cross-links, and metal complexation by multivalent ions tends to salt-bridge the neighboring molecular components together. As a result, one might think that capsules should exhibit very little metal-binding selectivity. Nevertheless, Mittelmann and Geesey[69] found that purified capsular material from a freshwater bacterium had a high affinity for copper. Certain metals (e.g., Mg, Ca, Fe, and K) can even influence the production of exopolymers, to which they will eventually bind.[63,70,71]

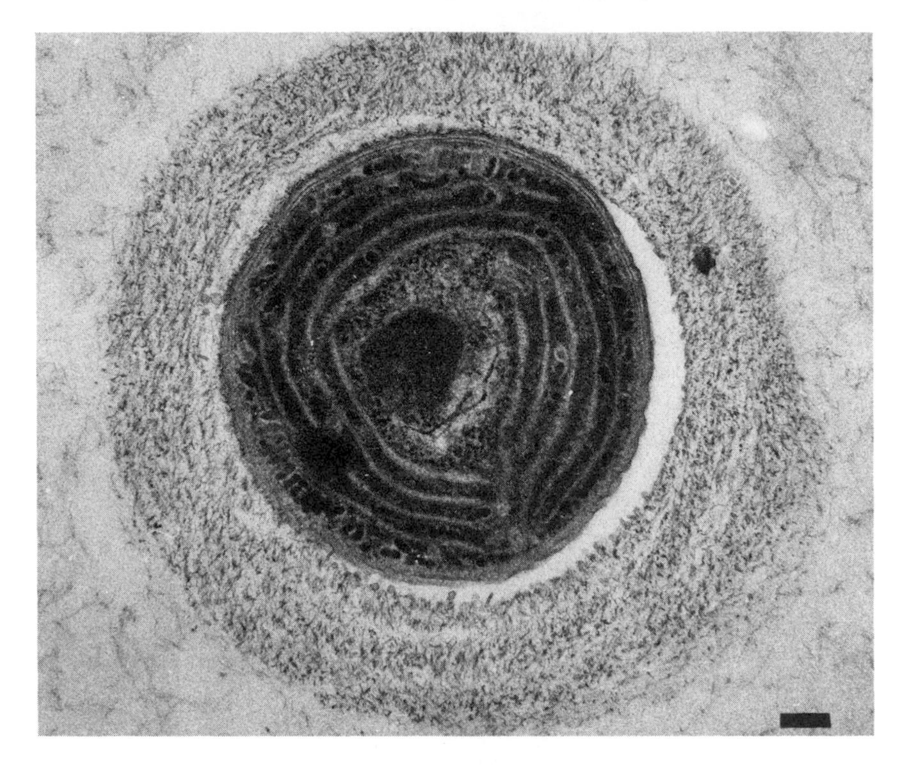

FIGURE 5. Thin section of the capsule on a cyanobacterium from a natural environment. Bar = 100 nm.

Metal complexation by capsules may play an important role in regulating the toxic heavy-metal concentration in the environment.[72-74] Bacterial capsules can act as effective modulators of metal ion concentrations at the cell surface, either by scavenging metals from solution when their concentrations are low and also by serving as an impermeable barrier when heavy-metal concentrations exist at toxic levels in the surrounding environment. This can affect the ecological and geochemical processes of the environment. In addition, for industrial surfaces, capsular metal binding is known to also contribute to corrosion reactions.[63]

c. Sheaths Sheaths are relatively rare in the bacterial world when compared to capsules. However, certain bacterial groups, like filamentous cyanobacteria, *Beggiatoa*, *Sphaerotilus*, and *Leptothrix*, commonly possess them as a loose arrangement of homo- or heteropolymers that encase chains of cells (Figure 6).[1,6] They resemble hollow cylinders when they are devoid of cells. The sheaths of *Leptothrix* and *Sphaerotilus* are noted for their ability to oxidize and precipitate iron and manganese from solution.[75]

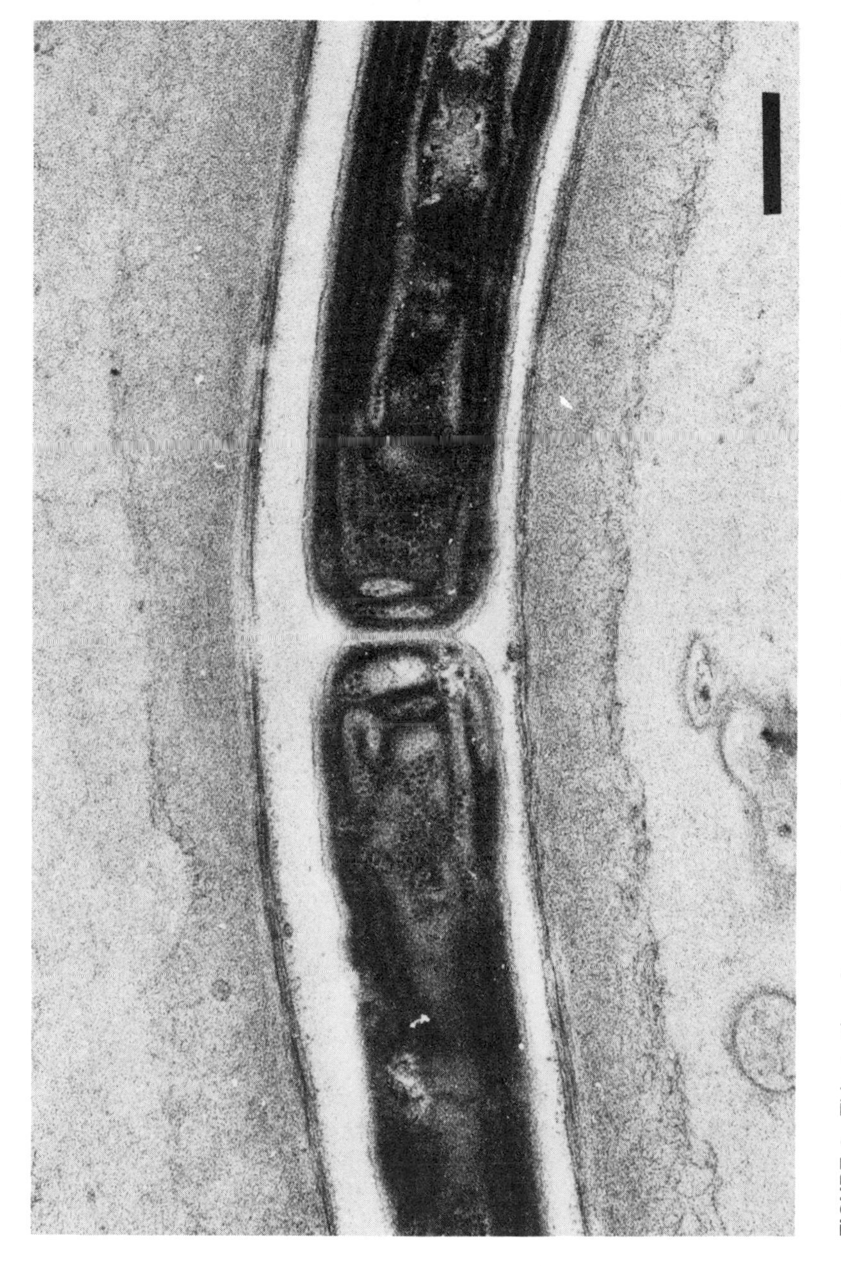

FIGURE 6. Thin section of a cyanobacterium from a freshwater environment possessing a fibrillar sheath. Bar = 500 nm.

Certain genera of the archaeobacteria, such as *Methanospirillum* and *Methanothrix,* possess more highly ordered sheaths than those mentioned above. These sheaths are composed of proteinaceous hoops which are stacked together to form long cylinders which encase a chain of cells.[76–78] In both cases, the methanogen sheath is the outermost surface layer and is in close contact with the underlying wall component. Generally, these sheaths can also bind a significant amount of metal.[13,114]

III. EXPERIMENTAL MODEL SYSTEMS USING WHOLE BACTERIA AND THEIR ISOLATED SURFACES TO STUDY METAL INTERACTIONS

Most of our discussion has concentrated on the structure and chemistry of bacterial envelope layers and their ability to complex metal ions. Intact bacteria are more complex than these layers in isolation, since they contain the cytoplasm which is an aqueous amalgam of metabolically active components. These components can be particulate (e.g., ribosomes) or soluble (e.g., cytosolic enzymes), and, at pH 6.8 to 7.0, most are anionic. Yet, most experiments using intact bacteria (e.g., see Mullen et al.[79]) or larger eucaryotic microorganisms (e.g., see Mullen et al.[80] and Mann et al.[81]) still point to cell surfaces as being most important for metal complexation and mineral development.

A. Laboratory Simulations

1. Simulations Using Intact Bacteria and Their Cell Walls

Low-temperature diagenetic processes that occur in natural sediments were experimentally simulated in the laboratory by suspending artificial metal-loaded *B. subtilis* cells in synthetic sediments and incubating at 100°C for up to 200 d.[82] The cells were loaded with metal prior to the simulations by suspending them in 5m*M* concentrations of either iron, copper, zinc, or uranium for 10 min at 22°C and removing all unbound metal by washing. The artificial sediment used consisted of "spec pure" calcite, crystalline quartz, or a mixture of the two. Elemental sulfur or crystalline magnetite was added to ensure that a relatively low Eh was maintained during the experiments.

Beveridge et al.[82] discovered the formation of a variety of metal phosphates; metal sulfides; and polymeric, metal-complexed, organic residues. They noted that the authigenic mineral phase which was formed commonly contained the metal which was originally loaded into the bacteria. In addition, the only source of phosphorus for phosphate minerals was the organophosphate of cellular components (e.g., nucleic acids, phospholipids, teichoic acid, etc.). Phosphate probably represented approximately 6% of the cell's dry weight. As a result, phosphate mineral formation appeared to correlate well with cellular decomposition.

In the presence of elemental sulfur, phosphate authigenesis did not occur, except at high pH values. Phosphate authigenesis was most pronounced during organic degradation of experimental runs containing uranium-loaded cells. Metal sulfides formed in the presence of elemental sulfur as a redox buffer. However, if the pH was elevated by the presence of calcite as a sediment, metal sulfide formation was blocked, and phosphate formation was promoted.

The results of Beveridge et al.[82] strongly indicate that bacterial cells are capable of actively nucleating mineralization during low-temperature diagenesis, and that cells can provide a major source of phosphorus for the production of phosphate minerals. At the same time, other metals previously bound by these cells may also become incorporated into authigenic minerals. The electron microscopic evidence showed that metal deposition initially occurred as small localized microcrystals. However, eventually these crystals increased in size and distribution.

In order to understand the geochemical processes involved in the cellular preservation of so-called "microfossils", Ferris et al.[83] conducted a series of laboratory experiments designed to simulate the fossilization of bacteria in sediment by iron and silica. They used *B. subtilis* cells loaded with iron to inhibit wall-degrading enzymes or autolysins[84,85] and control cells containing no added metals; these were then resuspended in an aqueous silica suspension and incubated at 70°C for periods up to 150 d. After just 30 to 60 d, the control cells completely lost their cell shape due to extensive autolytic activity. However, bacterial remains from the degradation of control cells did promote the mineralization of small silica crystallites after only 30 d of incubation. Further aging did not increase silicification in these controls.

In contrast, the iron-loaded cells retained their cell shape due to minimal autolytic activity. These results suggested that complexed iron increased both the autolytic and thermal stability of the constituent bacterial polymers. The authors also found that, although the precipitation of silica was slower with the iron-loaded bacteria, it eventually reached a stage comparable to the controls in 90 d. After this period, iron-loaded cells continued their slow deposition of silica until the cells were completely fossilized during the course of the experiment.

2. Simulations Using Bacterial Wall-Clay Composites

Natural aquatic sediments are much more complex than those of our initial laboratory simulations. They contain a much broader range of inorganic and organic particulates and solutes; it is difficult to extrapolate from simple laboratory simulations to these conditions. Clays are ubiquitous in soils and sediments, and are a logical addition in complexity to our bacterial simulations.[22,23,30,40,86] Since many metal-binding studies have been conducted on individual clay components (e.g., smectite and kaolinite) present in aquatic sediments,[87–90] there is also a broad base of knowledge in this area to compare

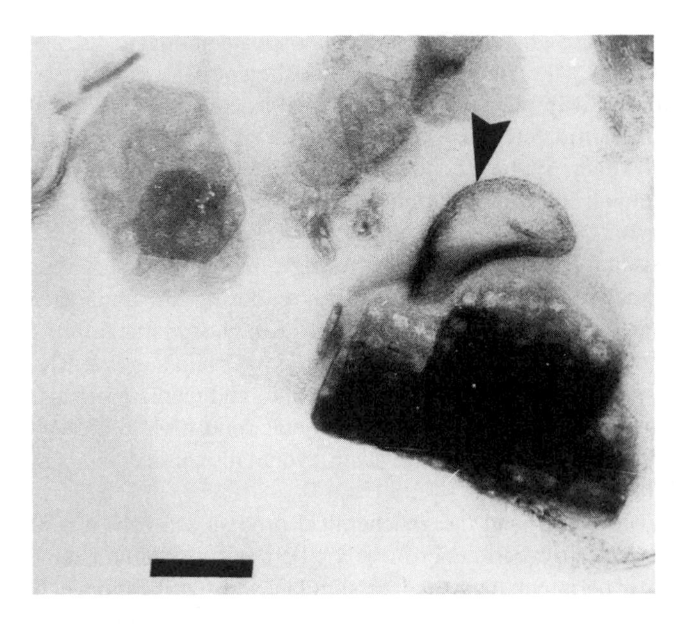

FIGURE 7. Thin section of an envelope (arrowhead) from *E. coli* attached to the edge of kaolinite. Bar = 100 nm.

with the bacterial studies. It was of interest to see how bacteria and clays interact with each other, and how their interaction might affect the known metal-binding capacity of each individual component.[91]

Walker et al.[91] used both Gram-positive *B. subtilis* walls and Gram-negative *E. coli* envelopes, which were individually mixed with fine-grained (<500 nm) smectite and kaolinite clays. They found that both *B. subtilis* walls and *E. coli* envelopes adsorbed to both clays, but each adhered more readily to smectite than to kaolinite. Overall, *B. subtilis* walls bound more efficiently to the clays than did *E. coli* envelopes. Electron microscopy revealed that the clays preferred an edge-on orientation with the bacterial walls (Figure 7). This obvious orientation preference indicated that the adsorption occurred between the positively charged edges of the clays and the negatively charged surfaces of the bacterial walls. However, the presence of multivalent metal ions increased the occurrence of planar surface orientations, which suggested that the cations were acting as salt-bridges between the bacteria and clay components.

Walker et al.[91] also determined the metal-binding capacity of isolated envelopes, walls, clays, and composite mixtures for Ag^{2+}, Cu^{2+}, Cd^{2+}, Ni^{2+}, Pb^{2+}, Zn^{2+}, and Cr^{3+}. Their results indicated that both bacterial components bound more metal than the clays, and that the Gram-positive walls adsorbed a greater quantity of metal than did the Gram-negative envelopes. More importantly, they found that the sum of the metal-binding capacity of each individual component exceeded that of their composites. This must be due to the elimination of many binding sites during the adhesion of the wall components to the clays.

Overall, the wall-clay and envelope-clay mixtures bound 20 to 90% less metal than did equal amounts of the individual components on a dry-weight basis. In all cases, except those of Cu and Ni envelope-smectite, the bacterial portion of the composite accounted for most of the metal-binding capacity. Flocs formed in many of the composite situations, especially in the walls-kaolinite system. Therefore, flocculation may have also reduced metal-binding sites on the individual components. Mutual flocculation of microorganisms and clays has been previously recognized in aquatic environments.[92] Generally, the results of Walker et al.[91] suggest that bacterial components in natural sediments are better sorbents for metals than for clays, and that the overall efficiency of both components is lowered when they are both present in the environment.

Flemming et al.[93] followed up these previous experiments by determining the conditions which could best remobilize the metals bound to the bacteria-clay aggregates. In this way, we can gain some insight into the nature and strength of the forces binding the metals to the bacteria-clay aggregates. Several remobilizing agents were tested at various concentrations. These included calcium nitrate (a competing salt), EDTA (a metal chelating agent), nitric acid (low pH), fulvic acid (an organic acid found in soils), and lysozyme (an enzyme capable of breaking down the peptidoglycan in cell walls). They discovered that under all conditions tested, less metal was resolubilized from the bacterial fraction than from the clay fraction. A clear pattern for the remobilization of each metal was not recognized, although the results were highly reproducible. These studies point out the many complexities of metal interactions in natural environments and also add an appropriate strong note of caution to the extrapolation of laboratory simulations to the natural environment. Ideally, detailed *in situ* studies are needed. Nevertheless, these present results strongly suggest that bacteria bind significantly more metal than clays under similar conditions, and that they are better able to keep their metal permanently bound. Therefore, bacterial components are a major metal-immobilizing agent within natural environments.

B. *In Situ* Field Studies

In studying natural environments, we invariably discover evidence of microbial metal binding and biomineralization, independent of whether or not we are examining planktonic or benthic populations. In this section, we examine some recent field studies on bacterial colonization, metal binding, and biomineralization in an attempt to emphasize just how prevalent microbial mineral development truly is in nature. We would also like to point out that the need for more detailed field studies cannot be underestimated, since these are worldwide processes of which relatively little is known.

1. Natural Biofilms in the Thames and Speed Rivers of Southwestern Ontario, Canada

Ferris et al.[94] examined microbial biofilms from submerged (10- to 20-cm water depth) rocks in rapidly flowing (10 cm per second) portions of the Thames River near London, Ontario and the Speed River near Guelph, Ontario. Biofilms were sampled by scraping 4.0-cm^2 areas of the surface of different types of rocks. Rock types included limestones, granite, gabbro, rhylolite, basalt, and quartz. However, only granite and limestone were sampled from both rivers. Additionally, freshly prepared blocks of commercial limestone and granite were submerged in the Speed River so that *in situ* microbial colonization could be followed through time. The granite blocks were polished on one half of the upper surface and abraded on the other half to produce microtextural differences between the two. These experimental blocks were sampled after 21 and 42 d of exposure in the Speed River.

Dense mucilaginous, greenish-brown, coatings covered the natural limestone surfaces from both rivers. In contrast, all other natural rock types examined appeared to be relatively clean to the naked eye. However, on scraping the surface of these rocks, a thin biofilm similar to that of the limestone was removed. Light and electron microscopic examination of these scrapings revealed abundant bacteria, cyanobacteria, filamentous eucaryotic algae, and diatoms (Figure 1). The bacteria were commonly encapsulated in microcolonies of morphologically identical cells. Therefore, these biofilms contained a complex community structure predominated by bacteria. Typically, the cyanobacterial and algal population densities were an order of magnitude less abundant than were corresponding bacterial population densities.

The Speed River counts were also generally lower than were the corresponding counts from the Thames River, and significant variations in biofilm population densities were observed between the different rock types. Again, limestone supported the highest microbial population densities (10^6 to 10^7 colony-forming units [CFU] per centimeter^2, where there were 10- to 100-fold more bacteria than in any other rock type examined. Quartz cobble supported the lowest overall population densities, with intermediate population densities recorded for the alkali feldspar/plagioclase-based rocks, including granite, rhyolite, gabbro, and basalt.[94] The results of this study suggest that there is an inverse relationship between biofilm cell density and substrate (mineral) hardness. The experimental granite and limestone blocks submerged in the Speed River exhibited very similar densities of adherent microorganisms after 21 d of *in situ* exposure. These observations were consistent with those of Mills and Maubrey,[9] in which they also found that freshly prepared specimens of quartz and calcite were colonized at approximately the same rate. However, after 42 d of exposure, the limestone blocks supported densities of bacteria, cyanobacteria, and algae greater than did the granite blocks. Similarly, the abraded halves of the granite blocks were colonized more heavily than were the polished halves. Consequently, the experimental results suggested that substrate hardness (and the ability of a surface to be abraded) has little effect on the actual microbial colonization rates, but does affect the final population density of epilithic microorganisms.

In summary, the experimental results provide strong evidence for a significant role of abrasive processes in determining the extent of microbial growth on mineral substrates. Rocks or minerals prone to abrasion appear to provide a better surface for microbial attachment and growth, particularly over longer periods of time.[94]

2. Moose Lake Watershed of Onaping, Ontario, Canada

Heavy-metal pollution is a major concern today, and, therefore, it is important for us to understand what happens to these metals in natural environments. Gaining this type of knowledge may help us in learning how to contain and possibly recycle these metals for future use. Thus, studies on metal uptake by natural adherent microbial biofilms and planktonic bacteria were carried out within acidic and neutral pH environments of the heavy metal polluted Moose Lake watershed located in Onaping, Ontario, northwest of Sudbury.[95] The biological and chemical oxidation of pyrrhotite and other metal sulfides present in the local mine tailings cause severe acidic (pH ~ 3) conditions in certain portions of the Moose Lake watershed. Acidic metal-laden mine drainage collects in the upper regions of the watershed where it is treated with a crushed limestone slurry to restore near neutral pH conditions and to reduce dissolved metals before the leachate reaches the lower region of the watershed. In this study,[95] four sample stations were utilized, two in acid waters (pH 3 to 3.5) and two in circumneutral waters (pH 6.5 to 6.9). Test and control settling plates were suspended *in situ* at the four sampling stations. Paired sets of test and control plates were removed and analyzed from each station after the first week and at approximately 5-week intervals thereafter, for a total *in situ* incubation time of 17 weeks. Concentrations of Mn, Fe, Co, Ni, and Cu were determined for the water column and for the biofilm and control settling plates.

Biofilm colonization data showed that both planktonic and adherent heterotrophic bacterial population densities rose from spring to autumn at all sample locations over the entire 17-week study period. Planktonic bacteria increased from 10^1 to 10^3 CFU per milliliter at all sites during the study period. However, biofilm growth increased more rapidly at all four sites over the first five weeks. The biofilm counts leveled off at 10^4 CFU per centimeter^2 at the neutral sites and 10^3 CFU per centimeter^2 at the acidic sites. Overall lower population densities at the two acidic sites suggested that bacterial growth may have been stressed by the low pH and high dissolved metal concentrations. Iron was the most concentrated metal species present at all four sites.[95] It accounted for up to 83 mol% of the five transition metals assayed. Nickel, manganese, cobalt, and copper showed consecutively lower concentration levels, with a 3- to 100-fold concentration decrease for all metals in solution at the neutral sites.

All metal species, except cobalt, were bound by the biofilms under acidic conditions at significantly higher amounts than were the corresponding controls. In general, metal adsorption was significantly greater at the neutral pH sites. In fact, microbial biofilm metal uptake at the neutral sites was enhanced by up to 12 orders of magnitude over acidic sites.[95] Thick coatings of iron oxide precipitated throughout the biofilms at these neutral sites and predominantly occurred as ferrihydrite on capsular extracellular polymers (Figure 8). The ferrihydrite exhibited a granular morphology and incorporated trace amounts of silica at neutral pH, whereas at acidic pH, it occurred as acicular crystallites containing traces of sulfur.

Metal cations are commonly adsorbed in smaller quantities at a low pH level due to a reduction of negatively charged sites on microbial surfaces where they can be neutralized by competing protons. In contrast, at neutral pH levels, metal adsorption is enhanced by an increase in the number of ionized acidic groups on microbial surfaces. Similarly, higher pH values also cause cationic colloidal elements (e.g., $Fe(OH)^{2+}.5H_2O$ or $Fe(OH)^{+}.4H_2O$) to form via hydrolysis of ferric iron.[96] These results indicate that natural metal-immobilizing biofilms are capable of removing significantly higher quantities of metal contaminants from solution than are inert surfaces. It is possible that these natural systems can be encouraged to help in the bioremediation of mine-tailing leachates.

3. Fayetteville Green Lake, New York, U.S.A.

Fayetteville Green Lake is an oligotrophic, alkaline lake that is unusual because of its deep meromictic nature and the presence of extensive carbonate bioherms (i.e., modern stromatolite/ thrombolite structures as shown in Figure 9). The history of research on this lake has been extensively reviewed by Thompson et al.,[60] along with a current detailed investigation of the lake's microbially-driven food chain and geochemistry. The lake chemistry is strongly influenced by groundwater input, since greater than 50% of the water supply comes from groundwater.[97] The water chemistry, particularly the high carbonate (3.57 to 7.43 mmol/l[98]); calcium (9.5 to 13.2 mmol/l[98]); sulfate (11.7 to 15.0 mmol/l[97]); and very low ammonium and iron concentrations (ppb range)[97] ensure that procaryotic phototrophs, such as the anaerobic purple and green sulfur bacteria and cyanobacterial picoplankton, are at an advantage over eucaryotic phototrophs. Procaryotic cells are extremely efficient scavengers of nutrients and trace metals in oligotrophic environments[99] and, therefore, they can successfully out-compete their eucaryotic counterparts in oligotrophic environments. As a result, procaryotic phototrophs are the dominant phytoplankton species in Green Lake.[60]

At the chemocline, which is the permanent transition zone between the oxic and anoxic regions of the lake, green and purple phototrophic sulfur bacteria thrive on the high amounts of sulfide emanating from the lower anoxic region of the lake (called the monimolimnion at approximately 18 to 53 m water

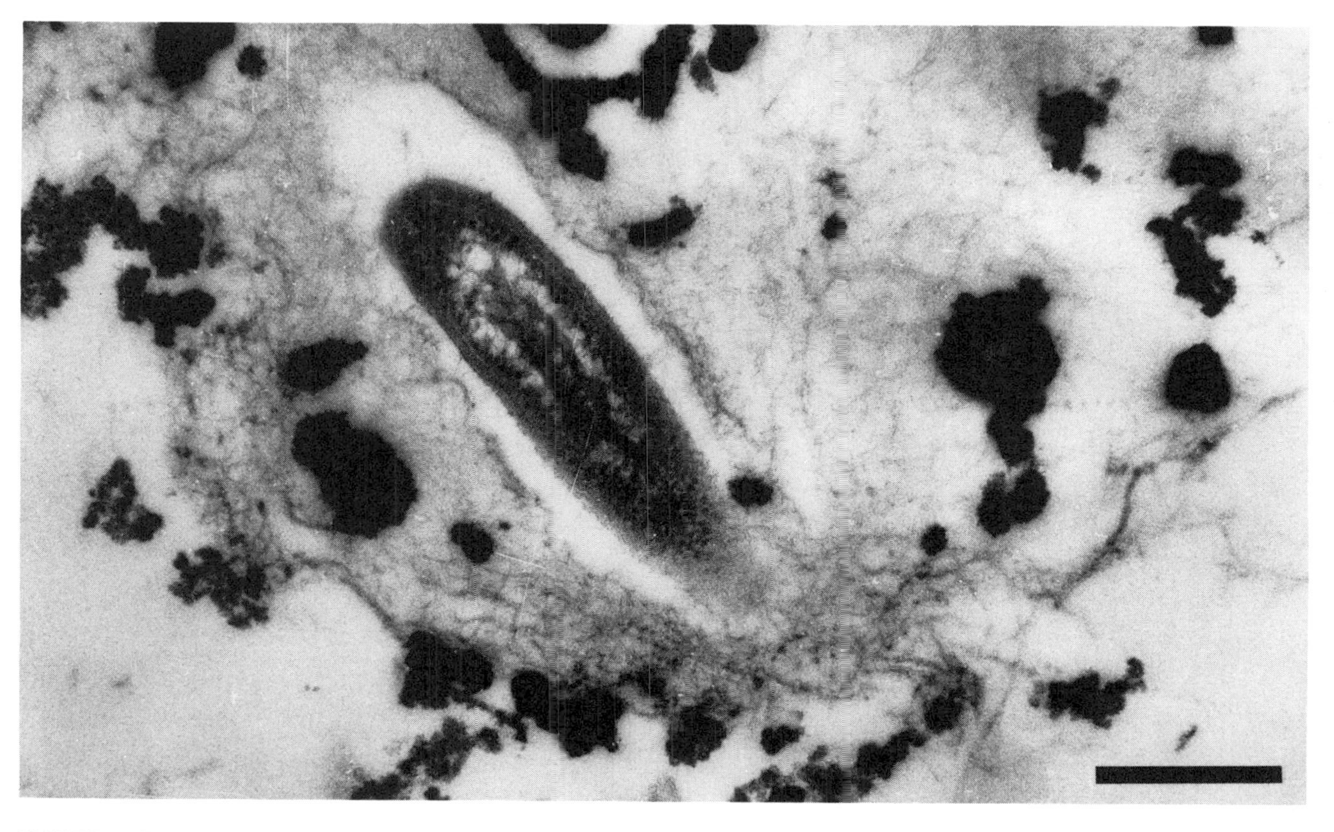

FIGURE 8. Thin section, *in situ*, of a Gram-negative bacterium from the Moose Lake region, northern Ontario, encased in a capsule containing iron oxides. Bar = 500 nm.

FIGURE 9. Underwater picture of the Fayetteville Green Lake, New York carbonate bioherms.

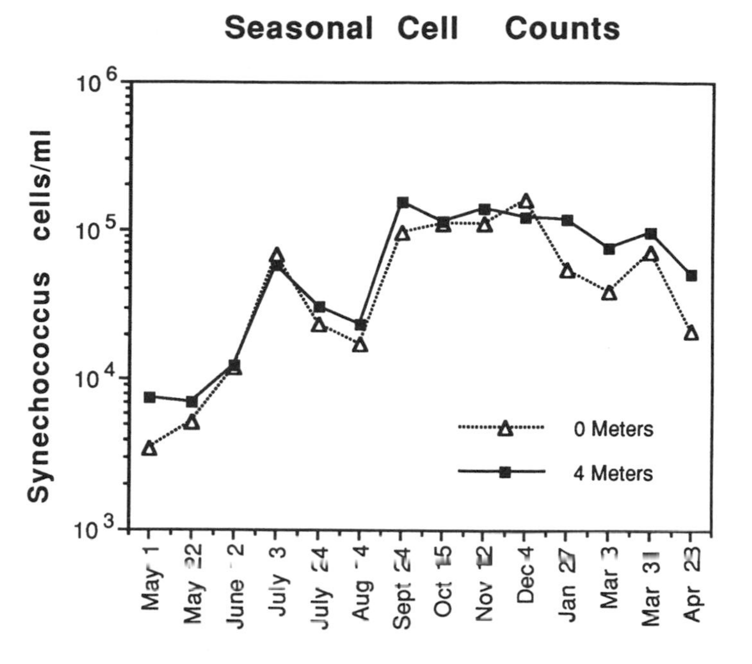

FIGURE 10. Seasonal growth of *Synechococcus* GL24 in Fayetteville Green Lake, NY.

depth). Sometimes the chemocline zone is referred to as the "bacterial plate" due to the extremely high concentration of bacteria in the water column at this depth (10^7 cells per milliliter of water). The bacterial population density at this layer is so thick that all sunlight is masked from penetrating deeper in the lake. Detailed studies of these purple and green sulfur bacteria in the lake are still lacking.

In contrast, a small (<1 μm diameter) unicellular cyanobacteria of the genus *Synechococcus* (designated strain GL24) appears to be the only significant phytoplankton species in the upper oxygenated mixolimnion.[60] It appears to colonize any hard substrate within the lake and is invariably found in high concentration in association with the carbonate bioherm material when viewed by either light or electron microscopy. *Synechococcus* thrives throughout the oligotrophic mixolimnion of Green Lake from its bloom in the water column during late spring to the late autumn. Its growth is seasonal, with high density peaks (10^5 cells per milliliter) during late June to early July and again in mid-to-late September (Figure 10). Its exponential growth rate occurs during late May to early June, which is, interestingly, the same time period when calcite mineralization is greatest in Green Lake.

Thompson and Ferris[59] experimentally demonstrated that *Synechococcus* was responsible for the major proportion of calcite mineralization in Green Lake, and, therefore, gives rise to both the annual whiting event and to the growth of carbonate bioherms within the lake. They discovered that *Synechococcus* actively induces epicellular calcite mineralization within the

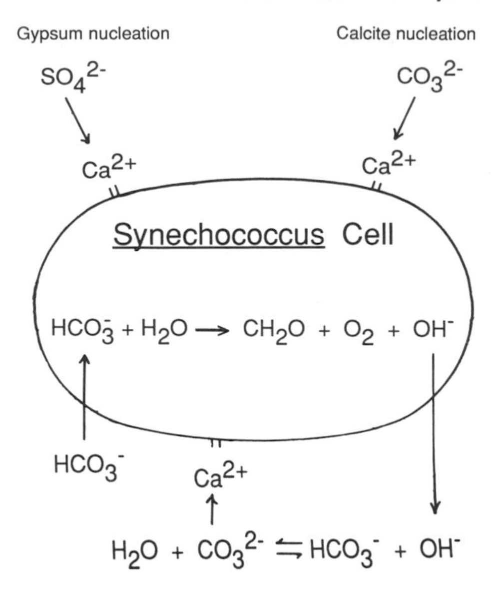

FIGURE 11. Model of the influence of photosynthesis (HCO_3^-/OH^-exchange) in *Synechococcus* GL24 on gypsum-calcite mineralization. Photosynthesis generates a microenvironment of OH^- around the cell, which raises its pH and promotes calcite mineralization vs. gypsum precipitation at more neutral pHs.

microenvironment surrounding each cell (i.e., the same unique envelope of water that we discussed previously in Section II.). Because of its photosynthetic metabolism and its use of HCO_3^- from the water as the primary source of carbon,[100,101] hydroxyl ions are pumped out of the cell and accumulate around the bacterium to increase its microenvironmental pH (i.e., HCO_3^- is "fixed" by the Calvin Benson cycle, and OH^- is released at 1 HCO_3^-:1 OH^-). Therefore, Synechococcus' photosynthetic alkalization actively drives the calcite precipitation (Figure 11), and consequently provides for the abundant marl sediment and stromatolite/thrombolite structures.[59,60]

In addition, laboratory simulations using natural lake water and *Synechococcus* GL24 also showed the epicellular mineralization of gypsum and magnesite (Figures 12,13). Epicellular gypsum precipitation was found to occur prior to calcite precipitation (especially at cooler water temperatures) and on "dark" control cells (i.e., no photosynthesis), which suggested that gypsum development was independent of photosynthetic alkalization.[59] Solubility products and mineral-saturation indices calculated from the experimental data agreed with the observed precipitation trend of gypsum, calcite, and magnesite (Figures 14,15). Initially, at time zero, the lake water plotted in the solid field with respect to gypsum and in the solution field with respect to calcite. This was a result of the starting metal concentrations and the circumneutral pH of the natural lake water (Figure 14; Table 2). The calculated saturation indices

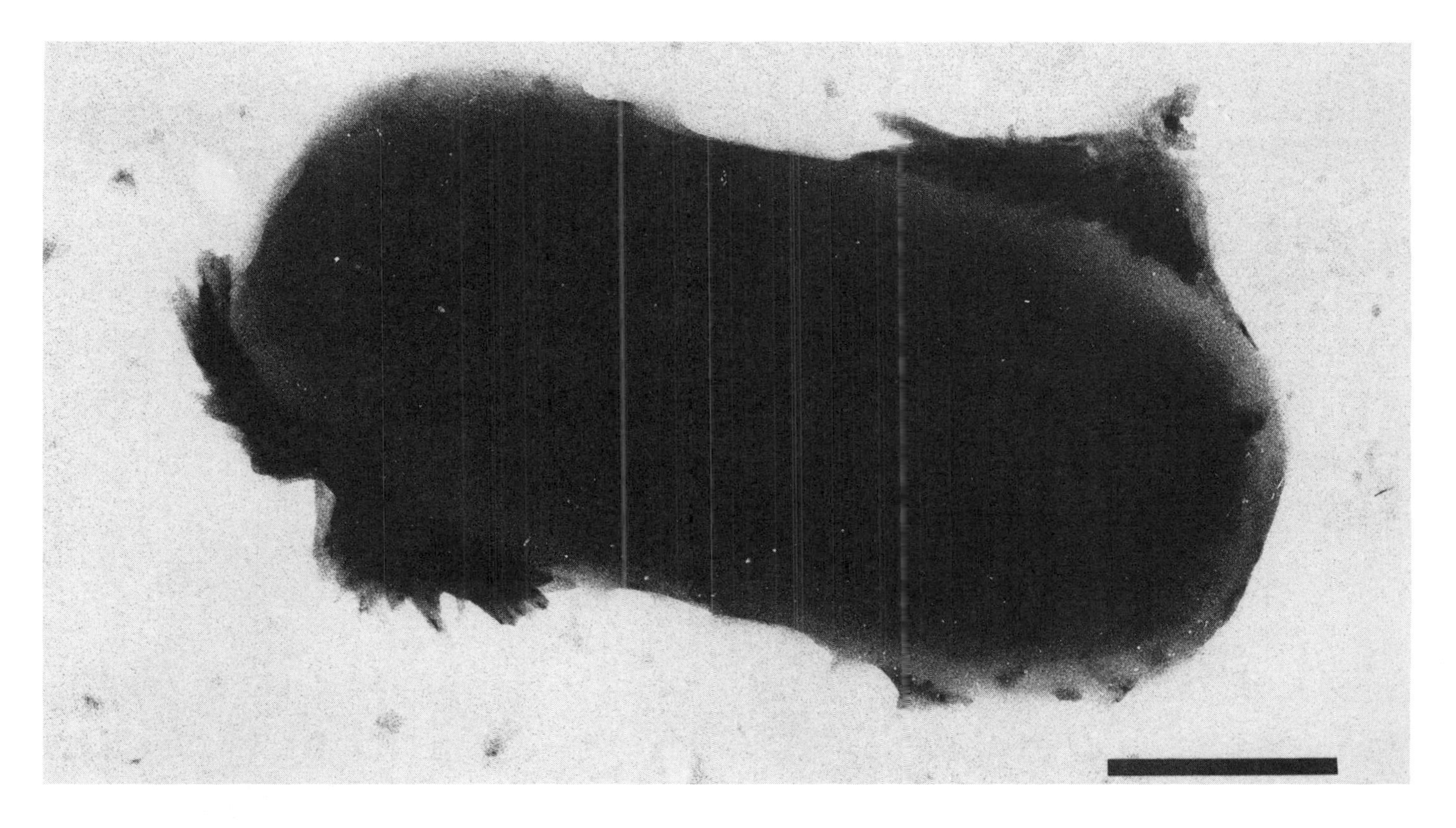

FIGURE 12. Whole mount of a *Synechococcus* GL24 cell showing abundant gypsum crystals on its surface. Bar = 500 nm.

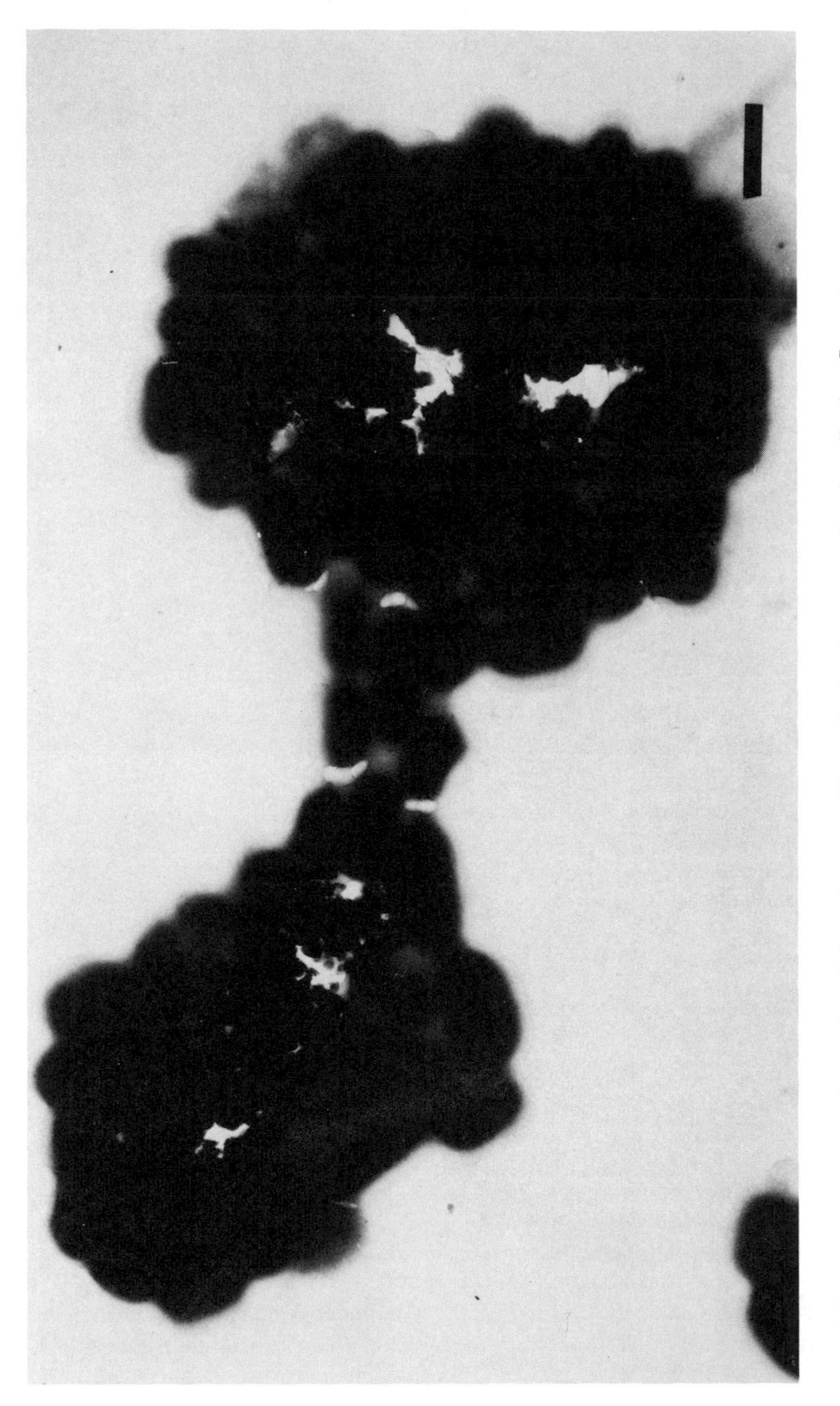

FIGURE 13. Whole mount of a number of *Synechococcus* GL24 cells showing magnesite mineralization. Bar = 1 µm.

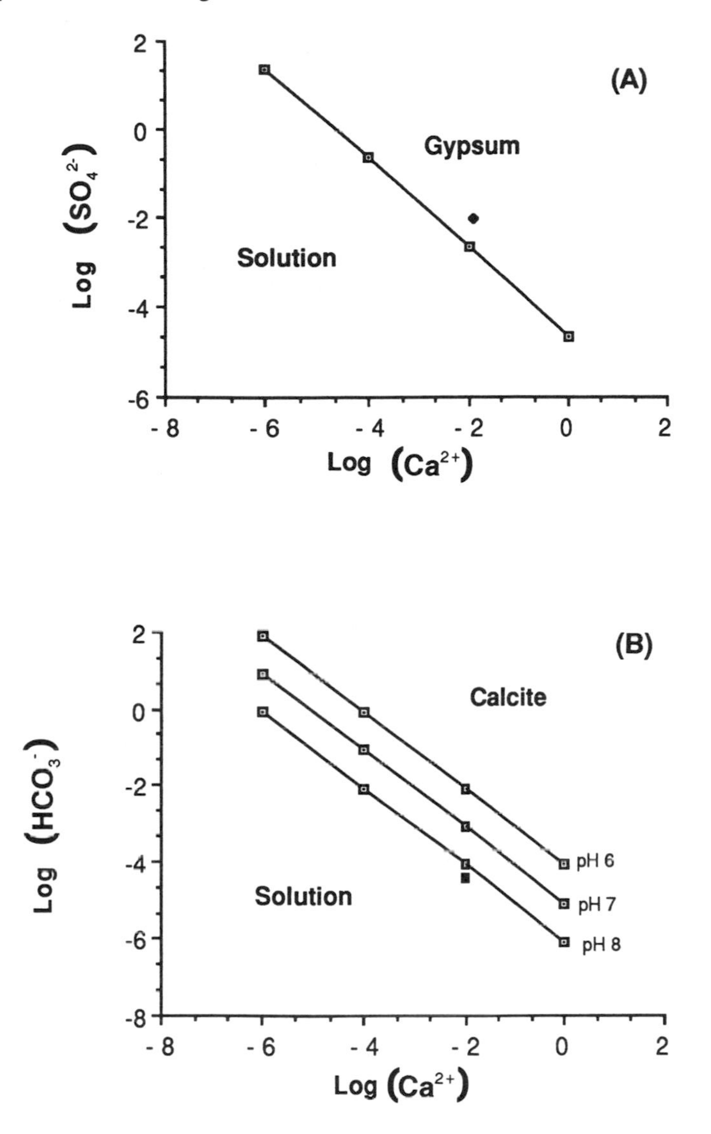

FIGURE 14. Plots to show the Green Lake chemical/mineralization saturation points for gypsum and calcite. The solid point (•) indicates the composition of lake water at start of experiment. For gypsum, it is in the solid field at the neutral pH of the Lake (only this pH is plotted), whereas the solid field of calcite is only approached at pHs >8.

indicated that the lake water was oversaturated with respect to gypsum (SI = 1.16) and calcite (SI = 2.73).[59] It appears that, under normal circumstances, the natural pH of the lake water is initially too low for calcite precipitation. Therefore, gypsum precipitates first on the cell surface, and, as warmth and

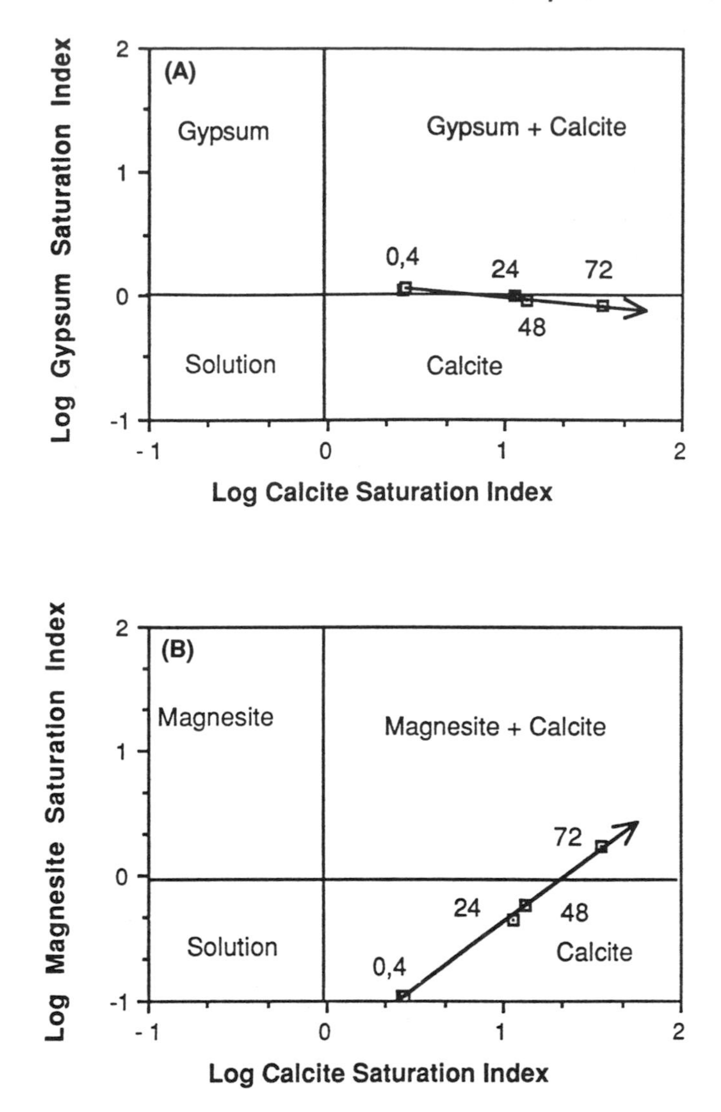

FIGURE 15. Green Lake $CaSO_4/CaCO_3$ and $MgCO_3/CaCO_3$ saturation indices. Numbers represent plots of sampling time intervals (0, 4, 24, 48, and 72 h) during calcification experiment. Arrows represent direction that saturation level was driven by microbial activity.

sunshine stimulate photosynthesis in the cell, alkalization promotes precipitation from gypsum to calcite over time (Figure 15). The gypsum, in turn, was eventually redissolved or replaced by calcite.

Certainly, temperature influences whether epicellular gypsum or calcite is formed. Observational evidence indicates that gypsum does precipitate in Green Lake during times of cooler water temperature, but this gypsum can be replaced by calcite during the warmer summer months. Gypsum precipitation seems to be

Table 2. Chemistry of Fayetteville Green Lake Water

Chemistry	May 15, 1988 (5-m water depth)
Ca^{2+}	11.39
Mg^{2+}	2.67
SO_4^{2-}	9.30
Alkalinity	3.20
Ionic Strength	55.20
pH	7.97

Note: Data in millimoles per liter.

From Thompson, J. B. and Ferris, F. G., *Geology,* 18, 995, 1990.

a more passive process not related to cellular metabolism, but more to a calcium-enriched cell surface and a high sulfate concentration in the water column.

Magnesite precipitation occurred after calcite was formed as the pH continued to rise in the external milieu of our experimental simulation. The presence of magnesite was first recognized after 72 h when the system became oversaturated with respect to soluble magnesite.[59] Magnesite was found only in association with large *Synechococcus* aggregates (Figure 13). The pH associated with these large cell aggregates may have been higher than that recorded in the experimental lake water, thereby resulting in the magnesite precipitation. Interestingly, magnesite has not been found in Green Lake, as the lake's natural pH level never exceeds pH 8.

These preliminary results on the microbial mineralization of gypsum and magnesite indicate that more field and laboratory investigations are warranted. Microbial magnesite precipitation is probably presently occurring in other natural aquatic environments with a high pH (8.5 to 10) that are also rich in magnesium. In fact, we are currently investigating such alkaline lakes in British Columbia.[102]

As mentioned (in Section II.2.a.) ongoing structural studies of *Synechococcus* GL24 have revealed a normal Gram-negative cell wall with an additional proteinaceous p6 S-layer (Figure 16).[62] The S-layer appears to act as a template capable of capturing and ordering Ca- and Mg-ions that promote mineral development. We have suggested that the extracellular S-layer may actually protect the cell from calcification occurring on the more delicate underlying cell envelope, thus preventing significant cellular damage. In this way, old calcified S-layer can be shed into the external environment, as it is simultaneously replaced by new S-layer material. Such a mechanism would allow for the growth and continued survival of *Synechococcus,* and, in a geochemical way, would encourage more calcification.

Synechococcus calcification in Fayetteville Green Lake provides an excellent example of the ability of a microorganism to effect overt physical changes in its environment. The appearance and growth of the 10-m-high carbonate bioherms (Figure 9) and the total thickness of marl sediment (meters) in the lake provide strong evidence that bacterial life has played and will continue to play a major part in the formation of geologically significant structures.

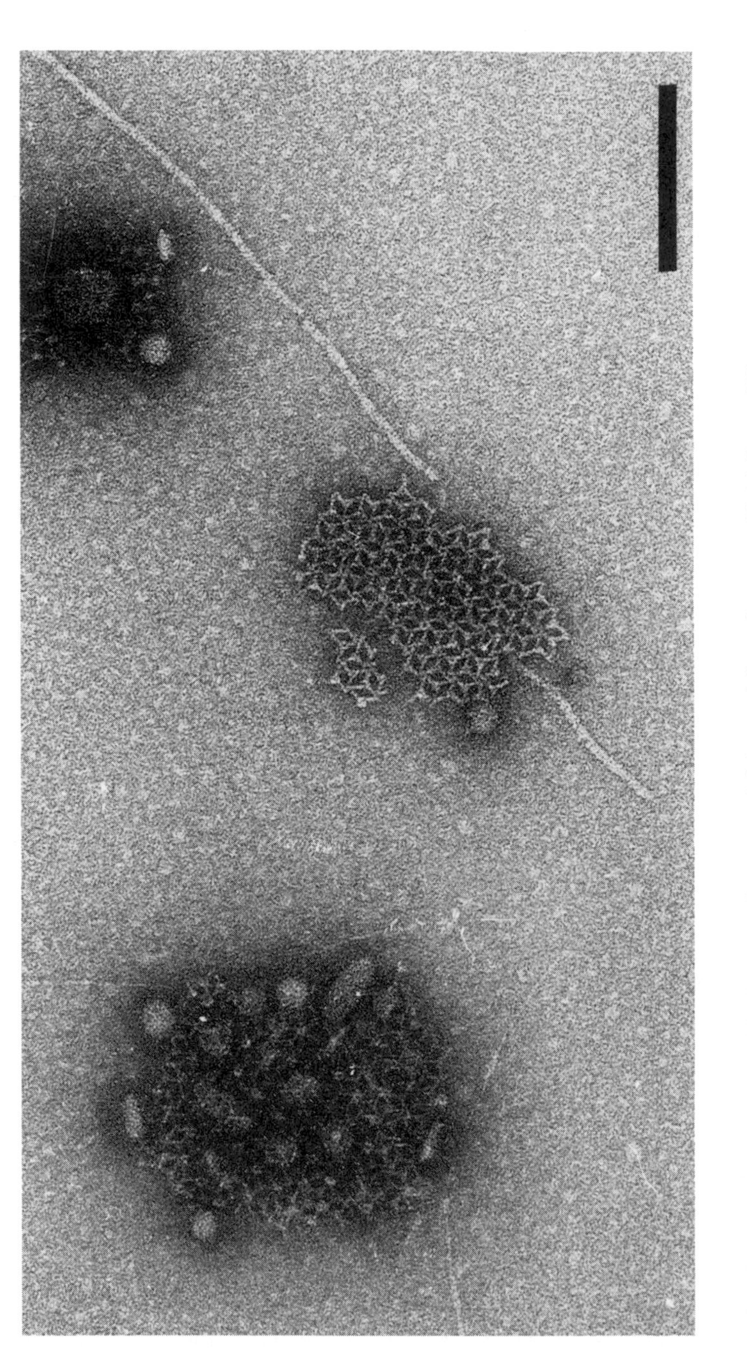

FIGURE 16. Negative stain of the hexagonally-arranged S-layer of *Synechococcus* GL24. Bar = 100 nm.

IV. METAL BINDING, MINERAL DEVELOPMENT, AND GLOBAL IMPLICATIONS

All bacterial surfaces normally contain a small amount of metal and this metal is, indeed, an integral part of the cell surface. Additionally, bacterial walls have been shown to tenaciously bind excess metallic ions in aqueous solutions.[21-26,86] In fact, bacteria actually have the greatest known capacity to bind and precipitate metals from aqueous solutions than has any other life form.[6]

Some bacterial walls seem to act as open-ion exchange resins,[26] whereas others seem more selective and exhibit an ability to partition certain metals.[22-24] "Steric fit" may be important in metal partitioning by bacterial surfaces, since ionic radius of a metal frequently determines whether or not complexation will occur.

Metal binding by bacterial surfaces is generally thought to be a passive phenomenon involving various strengths of electrostatic interaction between the anionic cell surface and cationic metals in solution. Accordingly, the bacteria do not have to be viable in many cases, only their surfaces or walls need to remain intact. Several studies have documented that dead bacteria can commonly bind higher quantities of metal than can living cells.[103-104] Not only will autolysin activity following the death of a cell increase the number of exposed metal binding sites in the cell wall, but actively respiring (living) bacteria actually pump protons into the wall by means of their energized membranes to compete with metal sorption.[105]

More often than not, the amount of metal adsorbed to the cell surface, in both living or dead cells, greatly exceeds the stoichiometry expected, based on the calculated number of reactive anionic sites within the cell wall.[21-23,39,40,86] Usually, precipitates can actually be visualized on the cell surface via electron microscopy because so much soluble metal has been precipitated (Figure 17). In time, these precipitates can eventually form crystalline metallic minerals. The initial metal adsorbed stoichiometrically neutralizes the chemically active anionic sites in the cell surface.[22,106] For example, available phosphate groups in teichoic acid are known to bind Na^+ or K^+ on a 1:1 basis. However, divalent cations like Mg^{2+} or Ca^{2+} bind on a 0.5 metal to 1.0 PO_4 basis, thereby acting as salt bridges between adjacent polymer strands.[27]

A two-step mechanism was invoked by Beveridge and Murray to explain the large quantities of metal bound by *B. subtilis* walls.[22] The first step, in time, consists of the stoichiometric interaction between metal ions in solution and the available active sites within the wall. The second step involves these stoichiometric bound metals as nucleation sites for additional metal adsorption from surrounding aquo-ions. This excess adsorption results in the formation of small, metal deposits that grow in time, using soluble anions from the external milieu as counter ions until the intramolecular spaces within the cell wall are filled. Eventually, these metal deposits are physically constrained by the lack of space within the wall framework, whereas those at the surface-fluid interface

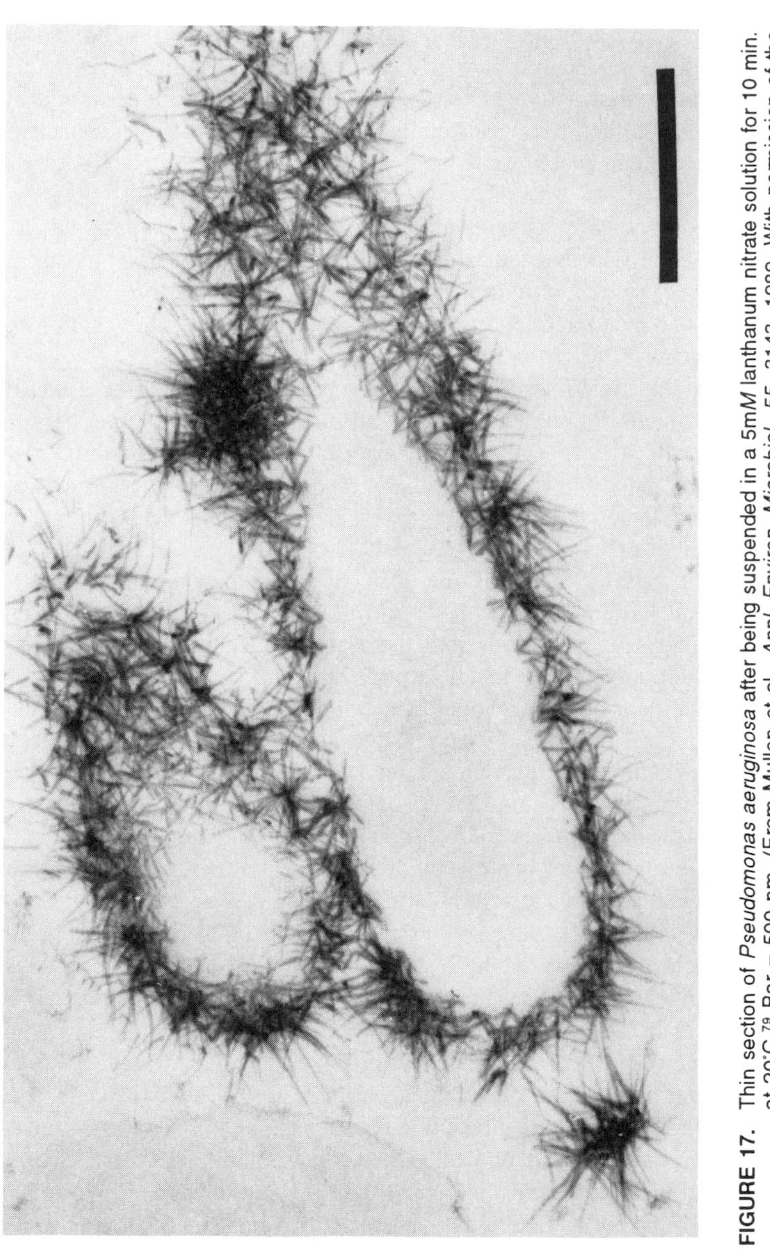

FIGURE 17. Thin section of *Pseudomonas aeruginosa* after being suspended in a 5m*M* lanthanum nitrate solution for 10 min. at 20°C.[79] Bar = 500 nm. (From Mullen et al., *Appl. Environ. Microbiol.*, 55, 3143, 1989. With permission of the American Society for Microbiology.)

continue to grow. In the end, a bacterial wall can contain copious amounts of metal, which can not easily be remobilized by competing agents.

Bacterial metal precipitates are initially hydrous, amorphous aggregates that resemble those seen in the early stages of mineral development. Consequently, in natural environmental settings more anhydrous, crystalline minerals should not be long in forming. A wide variety of precipitates and minerals have been found associated with bacterial surfaces *in situ* in natural settings. In fact, some researchers have been able to follow lithification and observe the amorphous aggregates become less hydrous and more crystalline.[107] The recent *in situ* finding that, in lake sediments, the development of (Fe,Al)-silicates, or clays, is possible is most startling.[108] Laboratory simulations suggest it is true,[109] and the implications of bacterially-induced clay formation could be staggering. The large number of different precipitates and minerals formed by bacteria is quite remarkable,[110] and this large scope suggests that there are a wide and varied number of biomineralization processes. A thorough understanding of the mechanisms involved is hampered by the extreme smallness of the interacting bacterial surfaces and by the sheer variety of minerals formed.

Finally, mineral formation and development is largely influenced by surface-charge density, by the cellular "microenvironment" (e.g., the unique biomineralization of *Synechococcus* GL24) and the "interfacial effect". The cellular microenvironment may possess unique pH, Eh, or elemental composition that influences the formation of minerals. At the same time, bacterial surfaces are always in contact with their external aqueous milieu, and therefore, represent a solid-fluid boundary to provide an interfacial effect. Therefore, simple-solution chemistry does not always apply. As a result, activation energies are thermodynamically lowered, and endothermic chemical reactions are made more achievable. Cell-surface physicochemistry is very important and is responsible for how different cells interact with their external aqueous environment. The physicochemistry of the bacterial surface is actually an essential design feature of procaryotic life. Therefore, it is our view that bacteria can both stimulate and accelerate mineral authigenesis, given the proper biogeochemical environment.

The global cycling of metals and minerals by bacteria has been in operation since the origin of life some 3.8 billion years ago. If we integrate the production rate of organic material over time, we get a total recycled mass of organic material that approaches the mass of the earth.[111] By association,[112] the mass of inorganic metals recycled through this organic material over time approaches the mass of the earth's crust! Consequently, bacterial mineral precipitation is not a trivial affair, especially if we consider that bacteria occur in almost all of the highly fractal surfaces of the earth. Bacteria and minerals are intimately entwined, since bacteria have evolved most efficient systems for immobilizing and retaining metal from their environment.[110] To truly appreciate the range and diversity of mineral development in association with bacterial surfaces and its global importance, we need to initiate many more *in situ* studies on natural environments, including soils, sediments, rivers, lakes, seas, and oceans and their simulations.

ACKNOWLEDGMENTS

S. Schulze-Lam of T. J. Beveridge's laboratory kindly supplied Figures 1 and 16. The fundamental structural work by the authors reported in this chapter was made possible by a Medical Research Council of Canada Operating Grant to T. J. Beveridge. Our biogeochemical work is supported by an operating grant of Natural Science and Engineering Research Council of Canada (NSERC) and an Ontario Geological Survey Ministry of Northern Development and Mines Grant to T. J. Beveridge. The electron microscopy was performed in the NSERC Guelph Regional Scanning Transmission Electron Microscope Facility, which is partially supported by an infrastructure grant of NSERC.

REFERENCES

1. Beveridge, T. J., The bacterial surface: general considerations toward design and function, *Can. J. Microbiol.*, 34, 363, 1988.
2. Balkwill, D. L. and Ghiorse, W. C., Characterization of subsurface bacteria associated with two shallow aquifers in Oklahoma, *Appl. Environ. Microbiol.*, 50, 580, 1985.
3. Jannasch, H. W. and Taylor, C. D., Deep-sea microbiology, *Ann. Rev. Microbiol.*, 38, 487, 1984.
4. Padan, E., Adaptation of bacteria to external pH, in *Current Perspectives in Microbial Ecology*, Klug, M. J. and Reddy, C. A., Eds., Am. Soc. Microbiol., Washington, D.C., 1984, 49.
5. Friedmann, E. I. and Ocampo-Friedmann, R., Endolithic microorganisms in extreme dry environments: analysis of a lithobiontic microbial habitat, in *Current Perspectives in Microbial Ecology*, Klug, M. J. and Reddy, C. A., Eds., Am. Soc. Microbiol., Washington, D.C., 1984, 177.
6. Beveridge, T. J., Role of cellular design in bacterial metal accumulation and mineralization, *Ann. Rev. Microbiol.*, 43, 147, 1989.
7. Geesey, G. G., Richardson, W. T., Yeomans, H. G., Irvin, R. T., and Costerton, J. W., Microscopic examination of natural sessile bacterial populations from an alpine stream, *Can. J. Microbiol.*, 23, 1733, 1977.
8. Geesey, G. G., Mutch, R., Costerton, J. W., and Green, R. B., Sessile bacteria: an important component of the microbial population in small mountain streams, *Limnol. Oceanogr.*, 23, 1214, 1978.
9. Mills, A. L. and Maubrey, R., Effect of mineral composition on bacterial attachment to submerged rock surfaces, *Microb. Ecol.*, 7, 315, 1981.
10. Mills, A. L and Mallory, L. M., The community structure of sessile heterotrophic bacteria stressed by acid mine drainage, *Microb. Ecol.*, 14, 219, 1987.
11. Fletcher, M., Effect of solid surfaces on the activity of attached bacteria, in *Bacterial Adhesion*, Savage, D. C. and Fletcher, M., Eds., Plenum Press, New York, 1985, 339.
12. Purcell, E., Life at low Reynold's number, *Am. J. Phys.*, 45, 3, 1977.
13. McLean, R. J. C. and Beveridge, T. J., Metal-binding capacity of bacterial surfaces and their ability to form mineralized aggregates, in *Microbial Mineral Recovery*, Erlich, H. L. and Brierley, C. L., Eds., McGraw-Hill, New York, 1990, 185.

14. Woese, C. R., Archaebacteria, *Sci. Am.*, 244, 98, 1981.
15. Woese, C. R., Kandler, O., and Wheelis, M. L., Towards a natural system of organisms: proposal for the domains Archaea, Bacteria, and Eucarya, *Proc. Natl. Acad. Sci. U.S.A.*, 87, 4576, 1990.
16. Beveridge, T. J., Wall ultrastructure: how little we know, in *Antibiotic Inhibition of Bacterial Cell Surface Assembly and Function*, Actor, P., Daneo-Moore, L., Salton, M. R. J., and Shockman, G. D., Eds., Am. Soc. Microbiol., Washington, D.C., 1988, 1.
17. Beveridge, T. J., Ultrastructure, chemistry, and function of the bacterial wall, *Int. Rev. Cytol.*, 72, 229, 1981.
18. Beveridge, T. J. and Davies, J. A., Cellular responses of *Bacillus subtilis* and *Escherichia coli* to the Gram stain, *J. Bacteriol.*, 156, 846, 1983.
19. Davies, J. A., Anderson, G. K., Beveridge, T. J., and Clarke, H. C., Chemical mechanism of the Gram stain and synthesis of a new electron-opaque marker for electron microscopy which replaces the iodine mordant of the stain, *J. Bacteriol.*, 156, 837, 1983.
20. Salton, M. R. J., The relationship between the nature of the cell wall and the Gram stain, *J. Gen. Microbiol.*, 30, 223, 1963.
21. Beveridge, T. J., Forsberg, C. W., and Doyle, R. J., Major sites of metal binding in *Bacillus licheniformis* walls, *J. Bacteriol.*, 150, 1438, 1982.
22. Beveridge, T. J. and Murray, R. G. E., Uptake and retention of metals by cell walls of *Bacillus subtilis*, *J. Bacteriol.*, 127, 1502, 1976.
23. Beveridge, T. J. and Murray, R. G. E., Sites of metal deposition in the cell wall of *Bacillus subtilis*, *J. Bacteriol.*, 141, 876, 1980.
24. Doyle, R. J., Matthews, T. H., and Streips, U. N., Chemical basis for the selectivity of metal ions by the *Bacillus subtilis* wall, *J. Bacteriol.*, 143, 471, 1980.
25. Matthews, T. H., Doyle, R. J., and Streips, U. N., Contribution of peptidoglycan to the binding of metal ions by the cell wall of *Bacillus subtilis*, *Curr. Microbiol.*, 3, 51, 1979.
26. Marquis, R. E., Mayzel, K., and Carstensen, E. L., Cation exchange in cell walls of Gram-positive bacteria, *Can. J. Microbiol.*, 22, 975, 1976.
27. Beveridge, T. J., The immobilization of soluble metals by bacterial walls, *Biotechnol. Bioeng. Symp.*, 16, 127, 1986.
28. Hussey, H., Sueda, S., Cheah, S. C., and Baddiley, J., Control of teichoic acid synthesis in *Bacillus licheniformis* ATCC 9945, *Eur. J. Biochem.*, 82, 169, 1978.
29. Kruyssen, F. J., de Boer, W. R., and Wouters, J. T. M., Effect of carbon source and growth rate on cell wall composition of *Bacillus subtilis* subsp. *niger*, *J. Bacteriol.*, 144, 238, 1980.
30. Kruyssen, F. J., de Boer, W. R., and Wouters, J. T. M., Cell wall metabolism in *Bacillus subtilis* subsp. *niger*: effects of changes in phosphate supply to the culture, *J. Bacteriol.*, 146, 867, 1981.
31. Ward, J. B., Teichoic and teichuronic acids: biosynthesis, assembly, and location, *Microbiol. Rev.*, 45, 211, 1981.
32. Burdett, I. D. J., Structure, growth, and division of the *Bacillus subtilis* cell surface, *Can. J. Microbiol.*, 34, 373, 1988.
33. Rogers, H. J., Perkins, H. R., and Ward, J. B., *Microbial Cell Walls and Membranes*, Chapman and Hall/Methuen, New York, 1981, 564.
34. Lambert, P. A., Hancock, I. C., and Baddiley, J., Occurrence and function of membrane teichoic acids, *Biochim. Biophys. Acta*, 472, 1, 1977.

35. Lifely, M. R., Tarelli, E., and Baddiley, J., The teichuronic acid from the walls of *Bacillus licheniformis* ATCC 9945, *Biochem. J.*, 191, 305, 1980.
36. Hobot, J. A., Carlemalm, E., Villiger, W., and Kellenberger, E., Periplasmic gel: a new concept resulting from the reinvestigation of bacterial cell envelope ultrastructure by new methods, *J. Bacteriol.*, 160, 143, 1984.
37. Graham, L. L., Harris, R., Villiger, W., and Beveridge, T. J., Freeze-substitution of Gram-negative eubacteria: general morphology and envelope profiles, *J. Bacteriol.*, 173, 1623, 1991.
38. Graham, L. L., Beveridge, T. J., and Nanninga, N., Periplasmic space and the concept of the periplasm, *Trends Biochem. Sci.*, 16, 328, 1991.
39. Hoyle, B. and Beveridge, T. J., Binding of metallic ions to the outer membrane of *Escherichia coli*, *Appl. Environ. Microbiol.*, 46, 749, 1983.
40. Hoyle, B. and Beveridge, T. J., Metal binding by the peptidoglycan sacculus of *Escherichia coli* K-12, *Can. J. Microbiol.*, 30, 204, 1984.
41. Nikaido, H. and Vaara, M., Outer membrane, in *Escherichia coli and Salmonella typhimurium, Cellular and Molecular Biology*, Vol. 1, Neidhardt, F. C., Ed., Am. Soc. Microbiol., Washington, D.C., 1987, 7.
42. Lugtenberg, E. J. J. and Peters, R., Distribution of lipids in cytoplasmic and outer membranes of *Escherichia coli* K-12, *Biochim. Biophys. Acta*, 441, 38, 1976.
43. Funahara, Y. and Nikaido, H., Asymmetric localization of lipopolysaccharides on the outer membrane of *Salmonella typhimurium*, *J. Bacteriol.*, 141, 1463, 1980.
44. Lam, J. S., Lam, M. Y. C., MacDonald, L. A., and Hancock, R. E. W., Visualization of *Pseudomonas aeruginosa* O antigens by using a protein A-dextran-colloidal gold conjugate with both immunoglobulin G and immunoglobulin M monoclonal antibodies, *J. Bacteriol.*, 169, 3531, 1987.
45. Ferris, F. G. and Beveridge, T. J., Binding of a paramagnetic metal cation to *Escherichia coli* K-12 outer membrane vesicles, *FEMS Microbiol. Lett.*, 24, 43, 1984.
46. Ferris, F. G. and Beveridge, T. J., Site specificity of metallic ion binding in *Escherichia coli* K-12 lipopolysaccharide, *Can. J. Microbiol.*, 32, 52, 1986.
47. König, H., Archaeobacterial cell envelopes, *Can. J. Microbiol.*, 34, 395, 1988.
48. Steber, J. and Schleifer, K. H., *Halococcus morrhuae*: a sulfated heteropolysaccharide as the structural component of the bacterial cell wall, *Arch. Microbiol.*, 105, 173, 1975.
49. Wieland, F., Dompert, W., Bernhardt, G., and Sumper, M., Halobacterial glycoproteins saccharides contain covalently linked sulfate, *FEBS Lett.*, 120, 110, 1980.
50. Koval, S. F., Paracrystalline protein surface arrays on bacteria, *Can. J. Microbiol.*, 34, 407, 1988.
51. Koval, S. F. and Murray, R. G. E., The superficial protein arrays on bacteria, *Microbiol. Sci.*, 3, 357, 1986.
52. Sleytr, U. B. and Messner, P., Crystalline surface layers on bacteria, *Ann. Rev. Microbiol.*, 37, 311, 1983.
53. Sleytr, U. B. and Messner, P., Crystalline surface layers in procaryotes, *J. Bacteriol.*, 170, 2891, 1988.
54. Beveridge, T. J. and Murray, R. G. E., Dependence of the superficial layers of *Spirillum putridiconchylium* on Ca^{2+}, *Can. J. Microbiol.*, 22, 1233, 1976.
55. Beveridge, T. J. and Murray, R. G. E., Reassembly *in vitro* of the superficial wall components of *Spirillum putridiconchylium*, *J. Ultrastruct. Res.*, 55, 105, 1976.

56. Buckmire, F. L. A. and Murray, R. G. E., Studies on the cell wall of *Spirillum serpens* strain VHA: the substructure and *in vitro* assembly of the outer structured layer, *J. Bacteriol.*, 125, 290, 1976.
57. Beveridge, T. J., Surface arrays on the wall of *Sporosarcina ureae*, *J. Bacteriol.*, 139, 1039, 1979.
58. Beveridge, T. J., Mechanisms of the binding of metallic ions to bacterial walls and the possible impact on microbial ecology, in *Current Perspectives in Microbial Ecology*, Klug, M. J. and Reddy, C. A., Eds., Am. Soc. Microbiol., Washington, D.C., 1984, 601.
59. Thompson, J. B. and Ferris, F. G., Cyanobacterial precipitation of gypsum, calcite, and magnesite from natural alkaline lake water, *Geology*, 18, 995, 1990.
60. Thompson, J. B., Ferris, F. G., and Smith, D., Geomicrobiology and sedimentology of the mixolimnion and chemocline in Fayetteville Green Lake, New York, *Palaios*, 5, 52, 1990.
61. Thompson, J. B., Schultze, S., Beveridge, T. J., and Fyfe, W. S., Cyanobacterial model system for the biological origin of gypsum and carbonate minerals, in *Geoscience Research Grant Program Summary of Research 1989-1990*, Milne, V. G., Ed., Ministry of Northern Development and Mines, Toronto, Ontario, 1990, 47.
62. Schultze-Lam, S., Harauz, G., and Beveridge, T. J., Participation of a cyanobacterial S-layer in fine-grain mineral formation, *J. Bacteriol.*, 174, 7971, 1992.
63. Geesey, G. G. and Jang, L., Interactions between metals ions and capsular polymers, in *Metal Ions and Bacteria*, Beveridge, T. J. and Doyle, R. J., Eds., John Wiley & Sons, New York, 1989, 325.
64. Costerton, J. W., Irvin, R. T., and Cheng, K. J., The bacterial glycocalyx in nature and disease, *Ann. Rev. Microbiol.*, 35,299, 1981.
65. Whitfield, C., Bacterial extracellular polysaccharides, *Can. J. Microbiol.*, 34, 415, 1988.
66. Smiley, D. W. and Wilkinson, B. J., Survey of taurine uptake and metabolism in *Staphylococcus aureus*, *J. Gen. Microbiol.*, 129, 2421, 1983.
67. Sutherland, I. W., Biosynthesis and composition of Gram-negative bacterial extracellular and wall polysaccharides, *Ann. Rev. Microbiol.*, 39, 243, 1985.
68. Altman, E., Brisson, J. R., and Perry, M. B., Structural studies of the capsular polysaccharide from *Haemophilus pleuropneumoniae* serotype, *Biochem. Cell Biol.*, 64, 707, 1986.
69. Mittelmann, M. W. and Geesey, G. G., Copper-binding characteristics of exopolymers from a freshwater sediment bacterium, *Appl. Environ. Microbiol.*, 49, 846, 1985.
70. Wilkinson, J. F. and Stark, G. H., The synthesis of polysaccharide by washed suspensions of *Klebsiella aerogenes*, *Proc. R. Physical Soc., Edinburgh*, 25, 35, 1956.
71. Corpe, W., Factors influencing growth and polysaccharide formation by strains of *Chromobacterium violaceum*, *J. Bacteriol.*, 80, 1433, 1964.
72. Hsich, K. M., Lion, L. W., and Schuler, M. L., Bioreactor for the study of defined interactions of toxic metals and biofilms, *Appl. Environ. Microbiol.*, 50, 1155, 1985.
73. Bitton, G. and Friehofer, V., Influence of extracellular polysaccharide on the toxicity of copper and cadmium toward *Klebsiella aerogenes*, *Microb. Ecol.*, 4, 119, 1978.

74. Aislabie, J. and Loutit, M. W., Accumulation of Cr(III) by bacteria isolated from polluted sediments, *Mar. Environ. Res.*, 20, 221, 1986.
75. Ghiorse, W. C., Biology of iron and manganese-depositing bacteria, *Ann. Rev. Microbiol.*, 38, 515, 1984.
76. Beveridge, T. J., Patel, G. B., Harris, B. J., and Sprott, G. D., The ultrastructure of *Methanothrix concilii*, a mesophilic aceticlastic methanogen, *Can. J. Microbiol.*, 32, 703, 1986.
77. Beveridge, T. J., Harris, B. J., and Sprott, G. D., Septation and filament splitting in *Methanospirillum hungatei*, *Can. J. Microbiol.*, 33, 725, 1987.
78. Southam, G. and Beveridge, T. J., Characterization of novel, phenol-soluble polypeptides which confer rigidity to the sheath of *Methanospirillum hungatei* GP1, *J. Bacteriol.*, 174, 935, 1992.
79. Mullen, M. D., Wolf, D. C., Ferris, F. G., Beveridge, T. J., Flemming, C. A., and Bailey, G. W., Bacterial sorption of heavy metals, *Appl. Environ. Microbiol.*, 55, 3143, 1989.
80. Mullen, M. D., Wolf, D. C., Beveridge, T. J., and Bailey, G. W., Sorption of heavy metals by the soil fungi *Aspergillus niger* and *Mucor rouxii*, *Soil Biol. Biochem.* 24, 129, 1992.
81. Mann, H., Tazaki, K., Fyfe, W. S., Beveridge, T. J., and Humphrey, R., Cellular lepidocrocite precipitation and heavy metal sorption in *Euglena* sp. (unicellular alga): implications for biomineralization, *Chem. Geol.*, 63, 39, 1987.
82. Beveridge, T. J., Meloche, J. D., Fyfe, W. S., and Murray, R. G. E., Diagenesis of metals chemically complexed to bacteria: laboratory formation of metal phosphates, sulfides, and organic condensates in artificial sediments, *Appl. Environ. Microbiol.*, 45, 1094, 1983.
83. Ferris, F. G., Fyfe, W. S., and Beveridge, T. J., Metallic ion binding by *Bacillus subtilis*: implications for the fossilization of microorganisms, *Geology*, 16, 149, 1988.
84. Herbold, D. R. and Glaser, L., *Bacillus subtilis N*-acetylmuramic acid L-alanine amidase, *J. Biol. Chem.*, 250, 1676, 1975.
85. Leduc, M., Kasra, R., and van Heijenoort, J., Induction and control of the autolytic system of *Escherichia coli*, *J. Bacteriol.*, 152, 26, 1982.
86. Beveridge, T. J. and Koval, S. F., Binding of metals to cell envelopes of *Escherichia coli* K-12, *Appl. Environ. Microbiol.*, 42, 315, 1981.
87. Farrah, H. and Pickering, W. F., The sorption of copper species by clays. I. Kaolinite, *Aust. J. Chem.*, 29, 1167, 1976.
88. Farrah, H. and Pickering, W. F., Influence of clay-solute interactions on aqueous metal ion levels, *Water Air Soil Pollut.*, 8, 189, 1977.
89. Gupta, G. C. and Harrison, F. L., Effect of cations on copper adsorption by kaolin, *Water Air Soil Pollut.*, 15, 323, 1981.
90. Pickering, W. F., Copper retention by soil/sediment components, in *Copper in the Environment, Part I, Ecological Cycling*, Nriagu, J. O., Ed., John Wiley and Sons, New York, 1979, 217.
91. Walker, S. G., Flemming, C. A., Ferris, F. G., Beveridge, T. J., and Bailey, G. W., Physicochemical interaction of *Escherichia coli* cell envelopes and *Bacillus subtilis* cell walls with two clays and ability of the composite to immobilize heavy metals from solution, *Appl. Environ. Microbiol.*, 55, 2976, 1989.

92. Avnimelech, Y., Troeger, B. W., and Reed, L. W., Mutual flocculation of algae and clay: evidence and implications, *Science*, 216, 63, 1982.
93. Flemming, C. A., Ferris, F. G., Beveridge, T. J., and Bailey, G. W., Remobilization of toxic heavy metals adsorbed to bacterial wall-clay composites, *Appl. Environ. Microbiol.*, 56, 3191, 1990.
94. Ferris, F. G., Fyfe, W. S., Witten, T., Schultze, S., and Beveridge, T. J., Effect of mineral substrate hardness on the population density of epilithic microorganisms in two Ontario rivers, *Can. J. Microbiol.*, 35, 744, 1989.
95. Ferris, F. G., Schultze, S., Witten, T. C., Fyfe, W. S., and Beveridge, T. J., Metal interactions with microbial biofilms in acidic and neutral pH environments, *Appl. Environ. Microbiol.*, 55, 1249, 1989.
96. Cotton, F. A. and Wilkinson, G., *Advanced Inorganic Chemistry*, John Wiley & Sons, New York, 1972.
97. Brunskill, G. J. and Ludlam, S. D., Fayetteville Green Lake, New York I: physical and chemical limnology, *Limnol. Oceanogr.*, 14, 817, 1969.
98. Torgersen, T., Hammond, D. E., Clarke, W. B., and Peng, T.-H., Fayetteville Green Lake, New York: ^{3}H-^{3}He water mass ages and secondary chemical structure, *Limnol. Oceanogr.*, 26, 110, 1981.
99. Fogg, G. E., Picoplankton, in *Perspectives in Microbial Ecology*, Megusar, F. and Gantar, M., Eds., Slovene Soc. Microbiol., Ljubljana, Yugoslavia, 1986, 96.
100. Miller, A. G. and Colman, B., Evidence for HCO_3^- transport by the blue-green alga (cyanobacterium) *Coccochloris peniocystis*, *Plant Physiol.*, 65, 397, 1980.
101. Badger, M. R., Bassett, M., and Comins, H. N., A model for HCO_3^- accumulation and photosynthesis in the cyanobacterium *Synechococcus* sp., *Plant Physiol.*, 77, 465, 1985.
102. Thompson, J. B., Ferris, F. G., and Beveridge, T. J., unpublished data, 1992.
103. Kurek, E., Czaban, J., and Bollag, J. M., Sorption of cadmium by microorganisms in competition with other soil constituents, *Appl. Environ. Microbiol.*, 43, 1011, 1982.
104. DiSpirito, A. A., Talnagi, J. W., Jr., and Tuovinen, O. H., Accumulation and cellular distribution of uranium in *Thiobacillus ferroxidans*, *Arch. Microbiol.*, 135, 250, 1983.
105. Urrutia, M., Kemper, M., Doyle, R., and Beveridge, T. J., The membrane-induced proton motive force influences the metal binding ability of *Bacillus subtilis* cell walls, *Appl. Environ. Microbiol.*, 58, 3837, 1992.
106. Rudd, T., Sterritt, R. M., and Lester, J. N., Formation and conditional stability constants of complexes formed between heavy metals and bacterial extracellular polymers, *Water Res.*, 18, 379, 1984.
107. Ferris, F. G., Tazaki, K., and Fyfe, W. S., Iron oxides in acid mine drainage environment and their association with bacteria, *Chem. Geol.*, 74, 321, 1989.
108. Ferris, F. G., Fyfe, W. S., and Beveridge, T. J., Bacteria as nucleation sites for authigenic minerals in a metal-contaminated lake sediment, *Chem. Geol.*, 63, 225, 1987.
109. Urrutia, M. and Beveridge, T. J., Formation of fine-grain silicate minerals and metal precipitates by a bacterial surface (*Bacillus subtilis*) and the implications in the global cycling of silicon, *Chem. Geol.*, submitted.
110. Beveridge, T. J. and Doyle, R. J., *Metal Ions and Bacteria*, John Wiley & Sons, New York, 1989, 461.

111. Abelson, P., Some aspects of paleobiochemistry, *Ann. N.Y. Acad. Sci.*, 69, 276, 1957.
112. Trudinger, P. A. and Swaine, D. J., *Biogeochemical Cycling of Mineral-Forming Elements*, Elsevier, Amsterdam, 1979, 612.
113. Ferris, F. G. and Beveridge, T. J., Physicochemical roles of soluble metal cations in the outer membrane of *Escherichia coli* K-12, *Can. J. Microbiol.*, 32, 594, 1986.
114. Sprott, G. D. and Beveridge, T. J., unpublished data.

CHAPTER 4

Particle-Associated Transport of Pollutants in Subsurface Environments

Aaron L. Mills and James E. Saiers

TABLE OF CONTENTS

0-87371-678-7/93/$0.00+$.50

INTRODUCTION

As with surface-water systems, we most often think of the transport of pollutants through groundwater habitats as occurring due to dissolution of a pollutant in the water followed by transport of the solution. Many times this is the case, and a number of reasonable models dealing with the fate and transport of pollutants have used solution transport as their basic assumption. Transport of contaminants in solution is not the only way that such compounds are moved in the subsurface, however. In a fashion similar to that observed in surfacial systems, pollutants may also be translocated in the subsurface as particle-associated complexes. The extent of such transport is unknown at present, but a few cases of important pollutant migration as a result of what is often termed "facilitated transport" have been documented.

The term "facilitated transport" has been used in several ways. One group of investigators, including those who study the transport of viruses, uses the term to indicate the observed early breakthrough of colloidal particles as compared to so-called conservative tracers due to hydrodynamic dispersion (this phenomenon will be explained in detail later). Alternatively, other workers, particularly those interested in the transport of contaminant molecules, use the term to refer to breakthrough of the contaminant due to its attachment to a mobile colloid particle. The latter use implies that a contaminant is mobilized or that its transport is accelerated because it is associated with a mobile, rather than an immobile, particle. Furthermore, the contaminant may even move faster than predicted by the behavior of conservative tracers. The latter behavior (which is included within the first use of the term "facilitated transport") will be explained in detail in the section on transport of colloids. For purposes of this discussion, facilitated transport shall mean the enhancement of contaminant transport due to association of the contaminant with mobile particles.

Facilitated transport of pollutants in groundwaters has a great many similarities to particle-associated transport on the surface, but many differences also exist. The fundamental chemical laws are the same in both types of systems, but the actual chemistry of the subsurface environment is often sufficiently different from that of surface locations to foster a unique set of conditions within which those laws operate. As a result, transport in the subsurface may be affected differently by changes in chemical properties of the environment as compared to transport in surface systems.

Physical laws are also the same in both systems, but the fact that flow velocities and the ratio of fluid volume to solids volumes (and therefore solid-surface area) differ radically from those at the surface means that different hydraulic properties may control transport in the subsurface as opposed to transport in surface environments. Differences in the solid to solution ratio can also have profound effects on the chemical interactions of pollutants, mobile particles, and mineral grains of the porous medium.

It is the purpose of this chapter to point out the properties of facilitated transport in subsurface environment. By comparison with other chapters, concepts that pertain to both systems may be adapted from knowledge of the surface. Concepts that do not seem to bridge the two systems will be described herein, and phenomena which are not well understood will be pointed out so that they may be explored in future research.

Perhaps the most significant difference between surface and subsurface environments that affects the behavior of the two systems is the dominance of the subsurface environments by surfaces. A simple calculation illustrates. In surface fluvial environments with very heavy suspended loads, say 1000 mg l^{-1}, of silt-sized particles (assumed to be 0.02-mm diameter spheres), the particulate surface area would be 1.13×10^5 mm^2. In an aquifer of fine sand (0.2-mm spherical grains, porosity = 0.38), the total surface area in 1000 cm^3 (a volume equivalent to 1 liter of water) would be 1.86×10^7 mm^2. The aquifer medium, has 165 times more surface than the material suspended in the river water, even though the particle diameter in this example is an order of magnitude lower in the suspension, and an extremely high suspended load has been used for the illustration. To carry the point of the dominance of surfaces in the aquifer to its most logical conclusion, however, it is appropriate to note that for the particles used in this calculation, the total particle volume of the suspended silt is 377 mm^3, whereas the aquifer material has a particle volume of 620,000 mm^3. If the particle surface area to fluid volume ratio is calculated, the real importance of surfaces in the subsurface environment can be seen. The river suspension has 113 mm^2 ml^{-1} of fluid. The porous medium, however, has 48,900 mm^2 of particle surface per milliliter of fluid. That is a factor of 433 times more surface area per unit volume of fluid in the sand aquifer than in the silt suspension. Clearly, the interactions between the surfaces of particles is of primary chemical importance in the subsurface environment.

II. HISTORICAL BACKGROUND

Traditional approaches for simulating subsurface migration of dissolved contaminants have assumed that the contaminant is partitioned between the immobile phase of the porous media and the mobile phase of the groundwater. However, recent evidence suggests that suspended colloidal particles provide an additional mobile phase for contaminant migration. Colloids are inorganic or organic particles having a diameter less than 10 μm [1] which tend to remain in suspension even under static conditions. The lower size limit is even less distinct, and is that point at which large molecules begin to take on the physical behavior of particles. It seems that point might vary depending on the specific nature of the molecules involved. Examples of inorganic colloids include iron oxides, aluminum oxides, and amorphous silica, while examples of organic colloids include macromolecular components of dissolved organic carbon and microorganisms, such as viruses and bacteria.

The earliest investigations into colloidal transport phenomenon were largely motivated by the desire to develop a clearer understanding of processes governing waste-water filtration. Most often, non-Brownian particles (>3 μm) were used in these studies. For example, Borchardt and O'Melia[2] examined the migration of unflocculated algal suspensions (15 μm to 60 μm) through filter sand (0.316 mm to 0.560 mm). Despite the large size of the algae, significant numbers were found in the effluent, and removal efficiencies of under 10% were observed. In a similar study using various strains of radiolabeled algae, Ives [3] reported that removal efficiency was directly related to algal size and inversely related to porous media grain size and pore-water velocity.

In addition to organic suspended matter, several investigations addressed the transport of inorganic colloids. Eliassen[4] examined the transport characteristics of non-Brownian, ferric iron suspensions in quartz sand beds. Most of the introduced iron was found to be removed in the upper layers of the sand bed, and only small concentrations of the iron appeared in the effluent after a considerable filtration period. In a series of papers, Hunter and Alexander[5-7] examined the physical and chemical mechanisms controlling the transport of kaolinite suspensions through quartz sand columns. Their results showed that kaolinite deposition was directly related to electrolyte concentration and temperature and inversely proportional to porous media grain size and fluid velocity. The strong dependence of deposition on electrolyte concentration indicated that electrostatic forces were instrumental in influencing the transport behavior of the kaolinite clay particles.

Although these studies used materials that were representative of the natural environment, they were all conducted under very high flow rates. Filtration studies are commonly performed at flow rates as high as 500 cm h^{-1}, which is orders of magnitude greater than common groundwater velocities. These high flow rates would likely decrease the magnitude of colloidal deposition relative to experiments conducted under more moderate fluid velocities. Consequently, these experiments can not be used as representative analogs of colloidal transport behavior in the natural subsurface.

With the exception of the investigations by Hunter and Alexander,[5-7] the previously described studies were primarily designed so that physical mechanisms of transport and deposition could be assessed. A substantial amount of research has also been conducted which attempts to elucidate the effects of chemistry on colloidal deposition and transport (e.g., Matijevik et al.,[8] Ives and Gregory,[9] Fitzpatrick and Spielman[10]). Specifically, Kuo and Matijevic[11] used the packed-bed technique to analyze the interactions of submicrometer hematite colloids with stainless steel beads as a function of pH and electrolyte concentration. It was found that hematite adhesion occurred over a narrow pH range where the particles and steel collectors were oppositely charged. Desorption was strongly dependent on the pH and ionic strength of the rinse solution. However, even at optimum conditions for desorption, only a small fraction of the hematite was detached from the collectors.

Kallay and Matijevik[12] used the same experimental system to examine the kinetics of hematite removal. It was established that particle desorption proceeds by two processes characterized by two significantly different rate constants. Initial desorption occurred very quickly and was believed to be represented by particles that were deposited on smooth surfaces of the steel beads. Following this initially rapid desorption stage, particles were removed much more slowly. These slowly desorbed particles were thought to have been originally adhered to rough surfaces of the collectors.

Although these studies with model collectors represent a necessary first step in development of a comprehensive theory of colloidal transport, they must be augmented by further experimentation using natural aquifer material. Depositional behavior in model systems will be substantially different from that of natural systems. Among the potential sources of variation are differences in charge distribution, surface roughness, and collector shape.

Evidence has accumulated which indicates that mobile colloidal particles significantly affect the transport and fate of strongly sorbing contaminants such as radionuclides and heavy metals. In some cases, the conclusions relate to the potential significance of the mechanism because the studies were carried out in laboratory sand columns. On the other hand, some of the studies have documented transport of the nuclides and metals in excess of that predicted using solution-based models in real systems, notably nuclear waste disposal sites.

Champlin and Eichholz[13] reported that virtually all ^{86}Rb, ^{46}Sc, and ^{140}La passing through a sand bed were associated with particulate matter. This observation prompted the authors to suggest that the means of storage of radioactive waste be examined carefully in repositories with a free-particulate component because the particulates and nuclides might combine, resulting in transport of the radioactive ions far beyond the confines of the reservoir. These general results and conclusions were supported by Eichholz et al.[14] Champ[15] observed that as much as 46% of plutonium eluted from soil columns was associated with particulate matter having a diameter of over 0.45 μm. Torok[16] found that cesium migration through soil was controlled by adsorption onto clay particles followed by clay-particle transport. Again the implication is that transport of the nuclides occurred over and above that predicted on the basis of ionic interactions with the immobile particles within the system. In a study of the mobility of Pu in groundwater, Mahara and Matsuzuru[17] determined that the production of mobile Pu during underground migration was strongly influenced by the amount of suspended solids in the groundwater. Giblin et al.[18] concluded that U mobility was linked to the presence of colloidal forms containing the element. Those forms could vary with the Eh-pH regime of the environment.

Bates et al.[19] examined the differential transport of Np, Am, and Pu from waste glass under storage conditions. While the authors determined that Np transport was as a mobile ion, Am and Pu removed from the aqueous suspension occurred almost entirely as colloidal solids. These authors concluded that Np performance (i.e., the ability of the repository to contain the nuclide) appeared to be adequately treated with current assumptions of solubility-controlled transport. The authors felt, however, that an engineered barrier system for the repository should include strategies to deal with the colloidal transport of Pu and Am. Furthermore, the authors suggested that assessments of the ability of the environment near the repository to retard nuclide movement should consider colloid migration as well as solution transport.

Buddemeier and Hunt[20] examined groundwater collected at the Nevada Test Site, both from within a nuclear detonation cavity and from a point about 300 m outside of the cavity. They found that all the transition metals and lanthanide (Ce, Eu) radionuclides were associated with colloidal particles, and concluded that the transport outside of the cavity represented colloidal transport. These authors indicated that anionic and neutral forms of the elements would be the most mobile, but that cationic and colloidal forms were more susceptible to ion exchange or capture by immobile aquifer surfaces. Because all of the nuclides found outside the cavity were colloid-associated, that form of transport must be of some importance in the aquifer studied.

Penrose et al.[21] found that Pu and Am detected over 3390 m downgradient from their seepage source were largely associated with particulate material. Predictions based on these actinides binding to the immobile aquifer solid forecast that migration would be limited to under a few meters, but the potential effect of the mobile colloids was ignored.

The obvious important application of these studies is that waste-repository siting and management based on transport models that consider only solution-phase transport may be inadequate to properly retain the nuclear material. The potential for significant colloidal transport should be considered in all siting and management plans.

In addition to radionuclides, other materials have been shown to be transported in colloidal association. Magaritz et al.[22] found that the concentrations of Zn, Cu, and Ag were enriched up to an order of magnitude in mobile, suspended matter in groundwater in the coastal plain aquifer of Israel. These authors suggested that colloidal transport of the metals in the carbonate, organic, and oxide phases moved the elements from sewage effluents and agricultural activities to the groundwater.

Magee et al.[23] reported a reduction of the retardation factor (enhancement of transport) of phenanthrene by an average factor of 1.8 in the presence of soil-derived dissolved organic matter (DOM). The authors suggest that significant transport of phenanthrene might occur as a phenanthrene-DOM complex. In this case, the term "dissolved" was an operational definition based on the extraction procedure, but the authors suggested that the material was all <5000 Da.

Dextran, but not a commercially prepared humic acid or a soil extract, enhanced the mobility of hexachlorobenzene and pyrene.[24] In fact, in this case the humic acid and soil organics retarded (decreased) the transport. The mobility of anthracene was not enhanced by any of the organics. In contrast, polychlorinated biphenyl (PCB) mobility is suggested to be strongly enhanced by addition of organics. A number of studies have reported substantial sorption of PCB to organic colloids in aqueous systems (e.g., Brownawell and Farrington[25] and Dunnivant et al.[26]) demonstrated increased mobility of hexachlorobiphenyl in the presence of organics. Additionally, these authors determined that Cd was cotransported along with the hexachlorobiphenyl in association with the organics. In all of the aforementioned cases, the authors concede that the exact nature of the organic as a truly dissolved species, a colloid, or a mobile particle larger than a colloid is not known.

III. POROUS MEDIA

A. Physical Setting

Groundwater aquifers vary substantially, both in physical and chemical nature. Physically, they may be comprised of a massive solid object with fractures running throughout, or they may be a loose aggregation of single mineral grains of virtually any size and shape. The former is usually not considered as porous media, and modeling flow and transport through fractured rock is one of the more challenging exercises currently being undertaken. Porous media are usually thought of as a granular matrix, although porous rocks (sandstones, limestones, etc.) often are included in this grouping because the behavior of water in them is nearly identical to that in a matrix of unconsolidated grains. The size of the pores depends on the size of the grains and the degree of aggregation of the grains (as does soil), but in all cases they range from zero to the approximate diameter of the grains. Some may be even larger, in the case of aggregations or of burrows or root channels. These larger pores are often called macropores, and although they are somewhat rare in terms of the total volume of material, they can be of great importance in terms of the flow of water. As we shall see later, transport of contaminants facilitated by particle association often depends on macropores for maximum speed and distance of transport.

The total amount of pore space in porous media varies. Dense crystalline rocks have porosities of 0 to 5% and may be virtually impermeable. Fractures in the rock may increase the porosity to a maximum of about 10%, but the fractures may permit the penetration of water where none could pass in the absence of the fractures. Sandstones, limestones, and fractured basalts may have porosities as high as 30 to 50%. Unconsolidated materials containing clays may have porosities reaching 60 to 70%. In those cases, the size of the pores is often so small that water is prevented from flowing through them. In terms of aquifers commonly considered in contaminant transport situations, the media are dominated by fine to medium sands or silts. In such cases, porosities of 40% often provide a close approximation to reality.

The most important physical term for porous media is the hydraulic conductivity (K). This term is a constant of proportionality (with units of length·time^{-1}) that arises from Darcy's law. The hydraulic conductivity integrates the effects of the porosity of the medium and the pore diameters that control flow. Values of K can span 13 orders of magnitude; high values are reported for gravels,

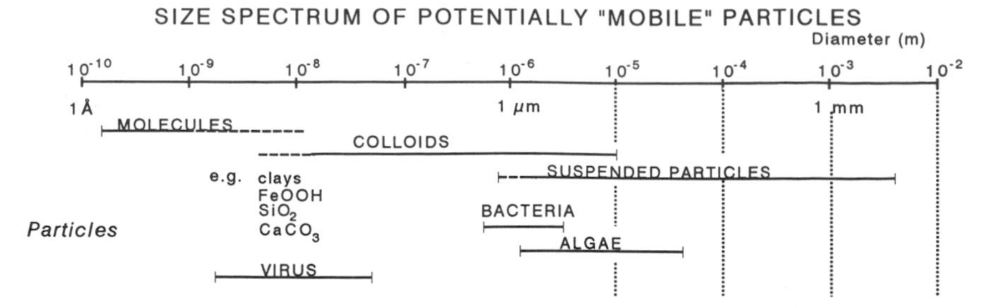

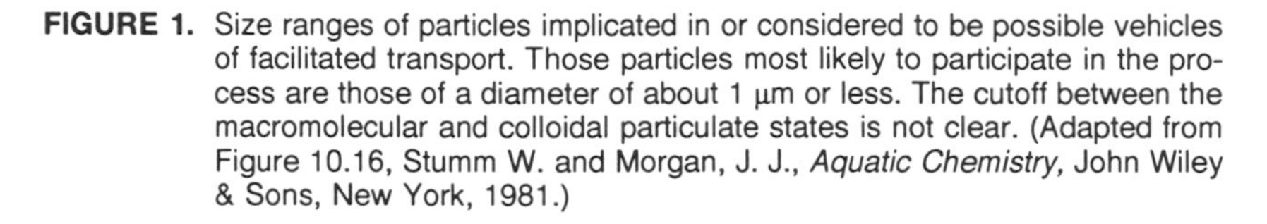

FIGURE 1. Size ranges of particles implicated in or considered to be possible vehicles of facilitated transport. Those particles most likely to participate in the process are those of a diameter of about 1 μm or less. The cutoff between the macromolecular and colloidal particulate states is not clear. (Adapted from Figure 10.16, Stumm W. and Morgan, J. J., *Aquatic Chemistry,* John Wiley & Sons, New York, 1981.)

whereas very low values are reported for unfractured metamorphic and igneous rock.

B. THE NATURE OF THE PARTICULATES

Particles that can facilitate transport of contaminants are of varying composition and origin. They may be inorganic materials or organic in nature, even to the point of being alive. Workers in the field often use the term "biocolloid" to refer to mobile particles that are of organic origin. It is, however, frequently unclear as to whether the term is used for all organically derived particles, or if it is reserved for the cells of bacteria, cells of very small eucaryotes, and viral particles. To eliminate confusion, we shall avoid use of the term altogether.

Particles that are mobile in subsurface environments are those that fall in the size range described by colloidal behavior. Stumm and Morgan[1] suggested that the upper limit on colloidal particles was approximately 10 μm (see Figure 1), but the more important particles in the subsurface are likely to be in the range of about 0.1 to 1 μm. Given the slow flow rates of most groundwaters, it is unlikely that even the particles in the 1 μm-size range will be maintained in suspension for long. Settling may accelerate transport in the downward direction, but it could slow the horizontal components of transport.

In many cases, the mobile particles are quite similar to the particles making up the porous medium. The colloids are often simply smaller-sized particles of the medium itself, produced by the slow, but continual, weathering of the medium grains. As described below, such particles may not be important in the facilitation of transport of contaminants. In many cases, the mobile particles are different from the porous medium itself in that the colloids have coatings of iron (sesquioxides) or organic matter. Most silicate clay minerals (e.g., kaolinite, muscovite, montmorillonite) are characterized by a negative charge that is weaker or stronger, depending on the nature of the mineral. The effect of coatings is to alter the magnitude of the negative charge, or in some extreme cases, to generate a net positive charge on the surface (as with FeOOH coating at alkaline pH).

IV. SURFACE PROPERTIES AND THEIR EFFECT ON CHEMICAL AND PHYSICAL INTERACTIONS

As pointed out above, aquifers are systems in which the physics and chemistry are dominated by the abundance of surfaces. This factor is particularly important in considering the movement of chemicals or particles through the porous medium. There would appear to be two major types of interactions between the immobile phase (mineral grains) and the mobile phase (pore fluid) of the system. The first is the group of attractions that are controlled by opposing electrostatic charges on the surfaces of the particles and associated with the contaminants or mobile particles. Such attractions include Van der Waals attractions, hydrogen bonding, dipole-dipole interactions, etc. The association of particles with the mineral grains is often described through the use of DLVO theory. Electrostatic attractions often allow multiple layers of molecules to sorb to the particle surface. The other type of attraction is the so-called hydrophobic effect, in which a nonsolvating material reaches a minimum free-energy state in its aqueous interaction by associating with a particle surface, thereby reducing the interfacial area exposed to the water. Both of these types of attractions (electrostatic and hydrophobic) are reversible and are often called physisorption, implying that no specific interaction between functional groups of the particle and the sorbing material exists.

In some cases, there is a specific interaction between the sorbent and the sorbate. This situation is usually referred to as chemisorption, and the associations are often irreversible without some major change in the chemical environment.

There are a number of ways to describe the strength of the associations; some (such as the Langmuir isotherm) are based in sound theory, while others (Freundlich isotherms and distribution coefficients) are empirical relationships. For the sake of simplicity in our discussions, we shall use the distribution (or partition) coefficient to describe the relative amount of contaminant that attaches to the mineral grains or mobile particles as opposed to that remaining in the water. For these purposes we define the distribution coefficient as:

$$m_{i,\,adsorbed} = K_d \cdot m_{i,\,solution} \tag{1}$$

or

$$K_d = \frac{\text{amount of contaminant adsorbed on solid}}{\text{amount of contaminant remaining in aqueous phase}} \tag{1a}$$

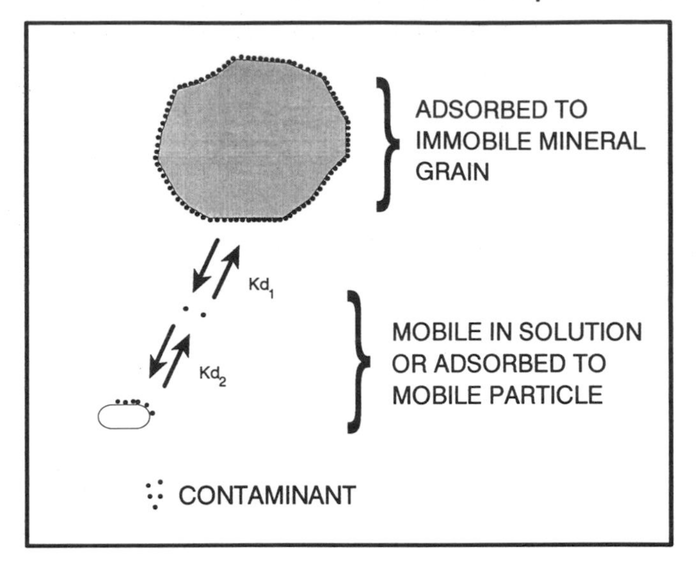

FIGURE 2. Competition between mobile and immobile particles for contaminant molecules. Assuming the K_d is always positive in the direction of movement from solution to a particle, then facilitated transport will be most important when $K_{d2} >> K_{d1} >> 0$, that is when the molecule partitions out of the water to a surface, and when the partition favors the mobile particle.

For compounds to be associated with mobile particles rather than immobile mineral grains, a relationship such as that shown in Figure 2 must be considered. Note that this is a general case for both physisorption and chemisorption. Irreversible attachment can be described simply by setting the respective K_d to very large numbers. (Note that the K_d approach has a problem in that it describes equilibrium behavior. It is well known that sorption in flowing liquids does not always reach equilibrium, and that sorption expressions which permit kinetic considerations must often be used to describe the phenomena quantitatively).

In the case shown in Figure 2, facilitated transport will occur only if K_{d1} is less than K_{d2}. Given an appropriate amount of time for equilibration, the contaminant will move from the immobile particle surface to the surface of the mobile particle for transport downgradient. If the mobile particles are simply smaller pieces of the porous medium grains, facilitated transport may occur, but its importance will be small because $K_{d1} \equiv K_{d2}$; thus the greater abundance of surface represented in the medium grains will hold the contaminant in place by means of mass action.

If on the other hand, the mobile particle-contaminant system has a higher K_d than the immobile particle-contaminant, facilitated transport can occur. The difference in the distribution coefficients can arise because of differences in the basic chemical makeup of the particles (mobile vs. immobile), or because the mobile particles have had their surfaces coated with material (such as iron or aluminum sesquioxide or organic matter) that enhances the K_d for the sorption. A conceptual of the conditions associated with facilitated transport at the pore scale is presented in Figure 3.

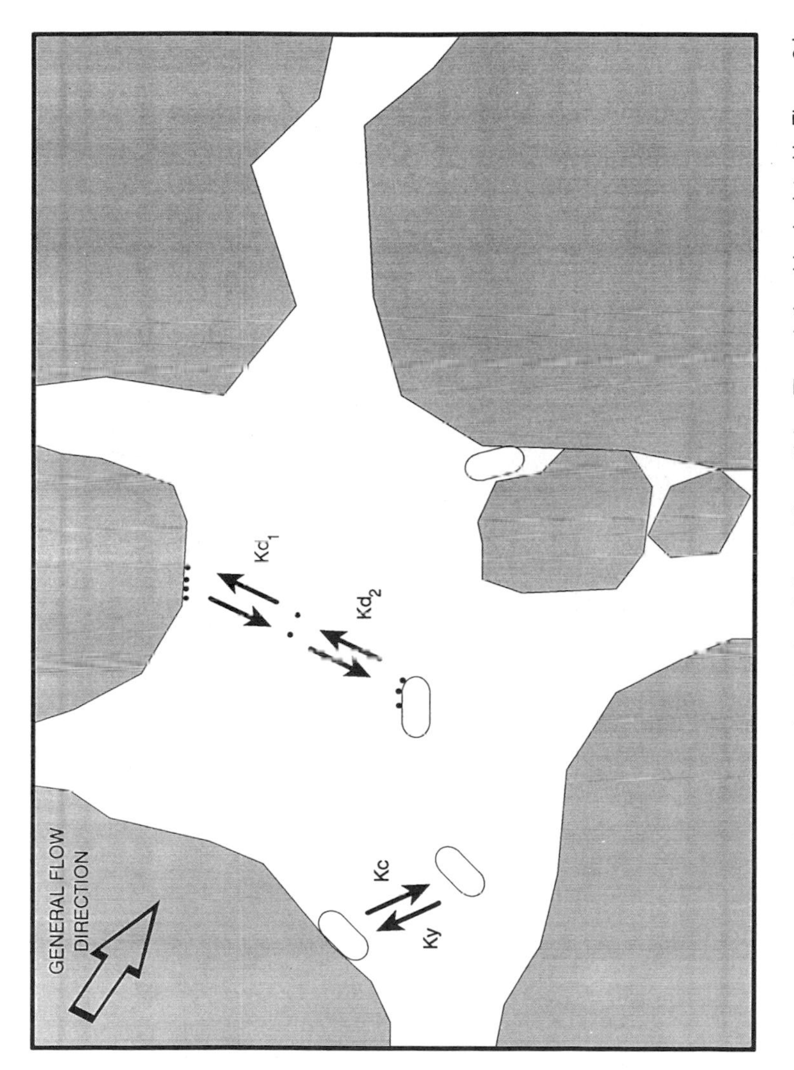

FIGURE 3. Facilitated transport of contaminants by mobile particles. The relationship depicted in Figure 2 is also shown here, as are deposition (described by K_c) and entrainment (described by K_y) of the particles on the immobile surfaces. Also shown is capture of the mobile particle in pores too narrow to allow passage. In mathematical models, this process is usually included in the deposition term.

V. TRANSPORT OF PARTICLES IN POROUS MEDIA

Quantitative descriptions of movement of particles in the subsurface are usually derived from some modification of the advection-dispersion equation [27-33].

$$\frac{\partial C}{\partial t} = D\frac{\partial^2 C}{\partial Z^2} - V\frac{\partial C}{\partial Z} - R \tag{2}$$

Equation 2 is presented here in a form that describes transport in one dimension. Two and three dimensional versions are commonly used (See Appendix X of Freeze and Cherry[34] for a full derivation), but the one-dimensional form allows for discussion of the concepts without the additional confusion of multidimensional notation. This equation contains terms that account for the movement of particles as they are carried along in the flowing fluid (advection), movement of particles by dispersion due to molecular diffusion, turbulence, tortuosity, and a term (R — for retardation or reaction) which is usually replaced with additional expressions that describe such phenomena as straining, adsorption/desorption, deposition (nonreversible sorption), etc. See Equations 3 to 7 for examples of "R" terms.

The advection term contains the variable V which is called by Freeze and Cherry[34] the "average linear velocity" of the water moving through some representative unit volume of porous medium. It is defined on the basis of Darcy's law and is the specific discharge, which has units of length per time, divided by the porosity. The use of this term implies that all of the flow properties, including tortuosity, are averaged as though all flow paths were exactly the same length. Obviously, this is not true, but for flow considerations, this macroscale integration of microscale properties is a reasonable approximation of the behavior of the system.

Dispersion is a more complex concept. It arises from the observation that if a quantity of solute molecules is inserted into a static solution at a single point, molecular diffusion (Brownian motion) will tend to turn the single point into a cloud of molecules with highest concentration at the center. If the point of molecules is inserted into a solution that is moving in some direction, the cloud will assume an elliptical rather than a circular shape, and the center of mass (highest concentration) will move in the direction of flow. Dispersion is not caused only by molecular diffusion; indeed, diffusion is too slow to account for most of the dispersion that is seen. Shear and turbulence in the mobile fluid are far more important than diffusion in determining the magnitude of dispersion. Thus, dispersion is caused primarily by differences in the flow velocities of the moving liquid. The result is often termed hydrodynamic dispersion. Additionally, the tortuous paths encountered in porous media greatly enhance hydrodynamic dispersion (See Figure 4). The term D in Equation 2 is called the dispersion coefficient.

Particles in suspension behave in much the same way as do solute molecules, although molecular diffusion is even less important at the particle scale.

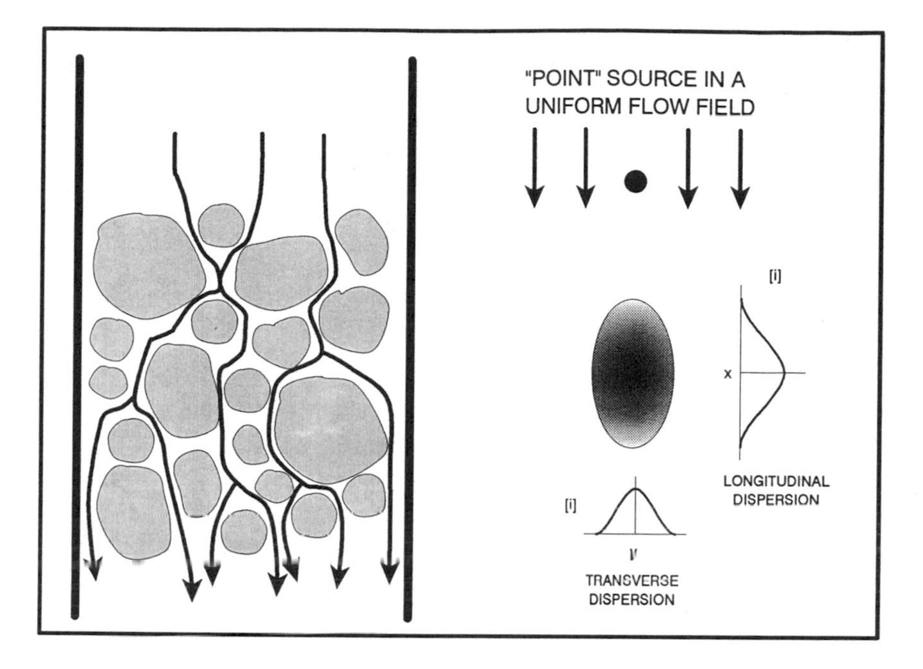

FIGURE 4. Appearance of hydrodynamic dispersion in a porous medium arising from tortuous pores. (Based on Figure 8.1 Wang, H. F. and Anderson, M. P., *Introduction to Ground Water Modeling: Finite Differences and Finite Element Methods,* W. H. Freeman, San Francisco, 1982.)

In this case, the tortuous paths exert a strong influence on the behavior of the particle cloud as it passes through the porous medium.

The transport behavior of solutes or particles is often described by so-called breakthrough curves. The name given to these curves arises from the viewpoint at which the behavior is measured. It is not always easy to measure concentrations of contaminant at all points along a flow path, so some points are selected for sampling. These points represent the ends of a flow path of some length defined as the distance from the point of origin to the sampling point. This situation is strictly analogous to viewing the system as a column of that same length, so that the contaminant is observed to "break through" the end of the flow path as a front of molecules or particles.

Breakthrough curves are constructed in several different ways, depending on the specific behavior to be observed. The x-axis is usually some type of measure of travel time, either in pure time units (e.g., hours, months, etc.) or in units representing the volume of material passing through the column or flow system (usually expressed as "pore", "bed", or "void" volumes, terms which all describe the volume occupied by the pore fluid under saturated conditions). Obviously, if one knows the flow velocity, pore volumes and travel time are directly interchangeable. The y-axis is expressed as concentration or as normalized concentration. In the latter case, the effluent concentration of whatever material is being examined (conservative tracer, contaminant, particles) is divided by the influent concentration (i.e., C/C_o). Thus, a breakthrough concentration equal to the influent concentration would be represented as 1.

Breakthrough curves have two basic shapes, depending on whether the contaminant is applied to the flowing groundwater as a pulse injection (as might be encountered in an accidental spill) or as a continuous injection (as might be encountered with a chronically leaking storage tank or with a leaking landfill). Pulse injections have the typical "chromatographic peak" shape seen in the right hand panels of Figure 5. The total mass passing through the column is represented by the area under the curve. If the y-axis is expressed as C/C_o, the maximum concentration would be 1 only if there is absolutely no dispersion (e.g., Figure 5, top right), even for "conservative" tracers. Furthermore, the peak would reach its maximum at exactly 1-pore volume. Continuous injections produce curves similar to those in the left-hand panels of Figure 5. In the ideal form of this situation, the concentration of the transported material increases until some steady state is reached. For conservative materials, the steady-state concentration will be equal to the influent concentration (C/C_o = 1). In the ideal case of no dispersion and perfectly conservative behavior, the concentration of tracer (contaminant, etc.) goes from 0 to 1 instantaneously, and the change will occur at exactly 1-pore volume. Figure 5 also shows the effects of dispersion on the breakthrough curves. Small amounts of dispersion cause a slight widening of the peaks in the pulse injections and a sloping of the breakthrough front in the continuous injection. Greater dispersion enhances these effects. It is important to note that dispersion causes breakthrough of tracer from the flow system prior to the time (1-pore volume) predicted by advection alone. Any phenomenon which enhances dispersion along the axis of flow (longitudinal dispersion) will enhance the early breakthrough of the subject material. As described earlier in this chapter, this accelerated breakthrough is what is termed facilitated transport by many investigators.

Equation 1 or variants thereof is used to describe transport that occurs in homogeneous porous media. Homogeneous media are, ideally, those with a uniform distribution of mineral grain sizes and pore diameters. Sorting is minimal in these systems. The most important point is that in the homogeneous case, preferred flow paths are absent. Preferred flow paths arise as fractures in rock, old worm burrows, root channels, or as veins of more transmissive (higher hydraulic conductivity) material embedded in the porous medium. Most of the mathematical models used to describe transport in porous media do not take heterogeneity, i.e., preferred flow paths, into account. For that reason, many of the models fail to predict breakthrough of subject material in advance of that calculated on the basis of advection-dispersion alone.[29]

The problems associated with accurate description of transport in heterogeneous media are amplified in the case of particle transport. Not only is a greater mass of particles transported in the presence of preferred flow paths, but transport occurs over a longer distance and more rapidly than is predicted by models based in the homogeneous case. This behavior is thought to arise because of size exclusion.

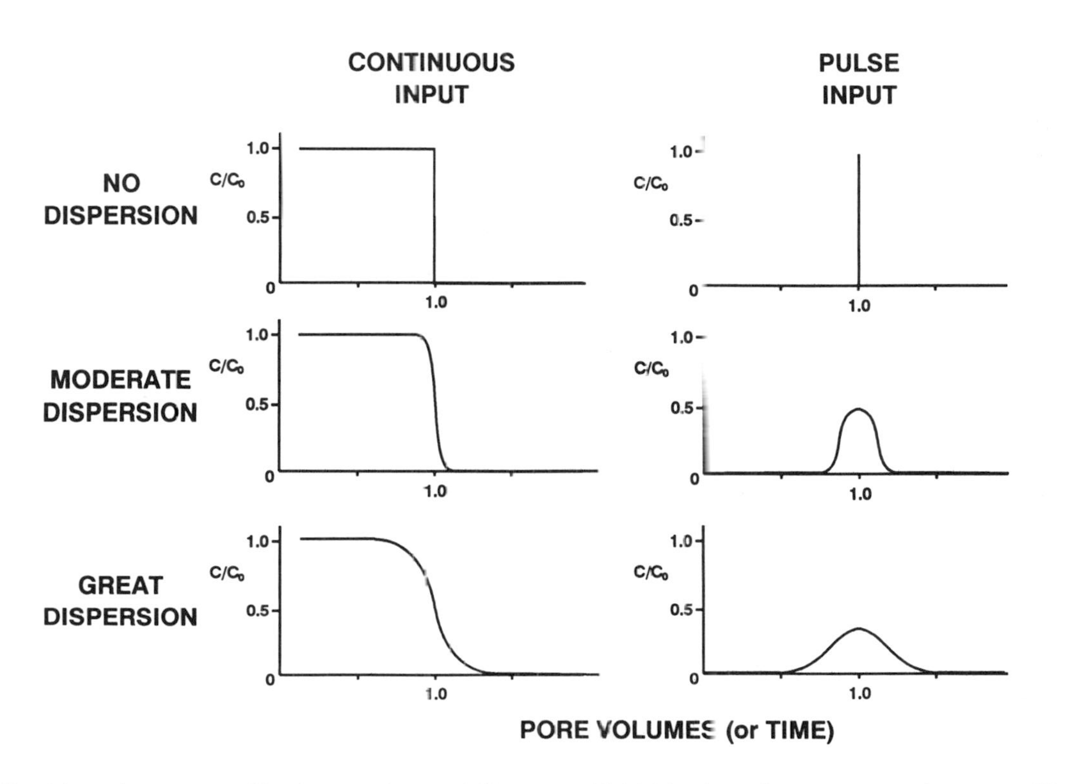

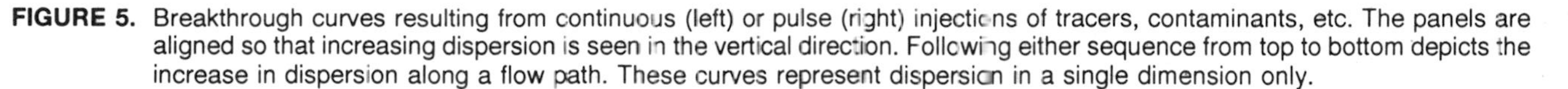

FIGURE 5. Breakthrough curves resulting from continuous (left) or pulse (right) injections of tracers, contaminants, etc. The panels are aligned so that increasing dispersion is seen in the vertical direction. Following either sequence from top to bottom depicts the increase in dispersion along a flow path. These curves represent dispersion in a single dimension only.

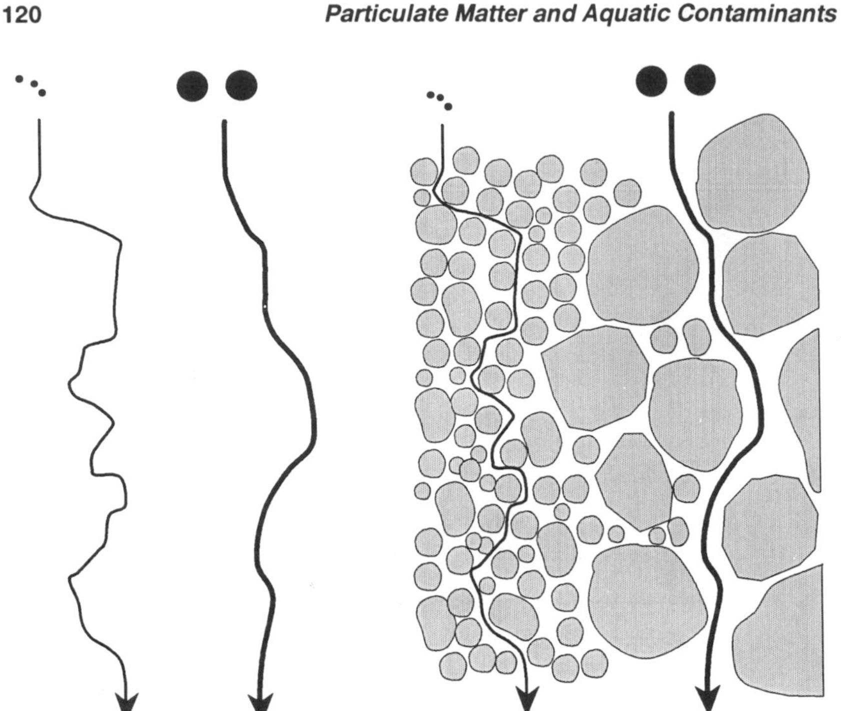

FIGURE 6. Size exclusion and its effect on travel time (i.e., path length) in heterogeneous porous media. Note that smaller particles could follow the exact path taken by the larger particles, but the probability of the small particles moving out of that path and into the smaller pores at some point along the way is high. Note that this diagram does not depict any type of retardation of the particles due to surface-surface interactions or pore clogging. In the original drawing, the path of the small particles represented a distance of 14.10 cm, whereas the path of the larger particles was 11.64 cm in length. Obviously the larger particles would leave the flow system ahead of the smaller particles in the finer matrix.

The size-exclusion principle simply states that larger bodies will be restricted to flow paths with the largest diameters. While the numbers of these paths may be small in any flow system, the paths are, theoretically, shorter in overall length than are the multitudinous paths defined by the less transmissive media (see Figure 6). Because the particles are not permitted to enter the small pores, transport is limited to the preferred flow paths. Total particle mass transported through these pores is typically very small, but it is greater than the mass that would be transported through the flow field in the absence of the heterogeneities. Thus, more particles are transported through the matrix at a faster velocity, and often for a longer distance than predicted by conventional application of the advection-dispersion model. It is this behavior that leads to accelerated transport of small amounts of materials in particulate form, or facilitated transport of otherwise immobile or slowly transported materials when associated with mobile colloids.

VI. MODELING FACILITATED TRANSPORT

Results of both laboratory and field studies demonstrate that, under many conditions, colloidal particles exhibit a high degree of mobility in groundwater environments. In an investigation of particle transport through a sandy aquifer, Harvey and George[35] reported that peak breakthrough of DAPI-stained bacteria occurred well in advance of bromide. Puls and Powell[36] showed that, under certain chemical conditions, radiolabeled Fe_2O_3 traveled almost conservatively through laboratory sand columns. Similarly, Saiers[37] observed that the colloidal silica transport through columns of quartz sand was nearly conservative and only slightly influenced by rate-limited sorption processes.

Recent theoretical treatments for evaluating contaminant migration in the presence of colloidal particles involve modifying the retardation factor of the advection-dispersion equation to account for partitioning of the contaminant between the aqueous, colloidal, and solid phases. Dunnivant et al.[26] used this approach to simulate the cotransport of both ionic and nonionic contaminants by naturally occurring DOC. Following similar lines, Enfield et al.[24] reparameterized the advection-dispersion equation to account for multiple mobile phases as well as for size-exclusion processes.

One limitation of these models is that they fail to consider changes in colloidal concentrations due to advective-dispersive transport of the colloids. Gradients in colloidal concentrations arising from advection and dispersion must be considered because they will influence partitioning of the contaminant between the mobile and immobile phases. The models implemented by Dunnivant et al.[26] and Enfield et al.[24] also do not account for changing colloidal concentrations resulting from sorptive reactions with the porous media. Rate limited adsorption has been shown to be important in governing the fate of both organic and inorganic colloidal particles.[32,38]

A number of models have been proposed for the transport of colloids. One of the more commonly cited formulations is that of Corapcioglu and Haridas.[27,28] This model suggests a basic theory for the transport of bacteria. It takes into account most of the fundamentals of particle transport, but also includes terms that allow for the growth and death of bacteria in the groundwater. The model provides a basic framework for homogeneous porous media, but does not account for preferential flow paths in the medium and cannot describe flows in structured soils.[39] The state variables treated in the theory are (1) C, the concentration of particles (in this case bacteria) in the flowing water; (2) C_f, the concentration of the growth-limiting substrate in the aqueous phase (important only for viable, growing bacteria); (3) S_f the mass of adsorbed substrate per unit mass of rock (important only if the transported particles are growing bacteria); and (4) σ, the volume of deposited particles (bacteria) per volume of bulk rock. The major assumptions incorporated into the theoretical framework are (1) growth of bacteria is governed by Monod kinetics; (2) bacterial death is proportional to bacterial concentration; (3) filtration is governed by a modified form of the kinetic expression of Mints[40]; (4) adsorption of substrate is governed by a Freundlich isotherm; and (5) a constant stoichiometric ratio between mass of substrate utilized and mass of bacteria produced exists. Note that assumption 3 is the only one necessary if the transported particles are not growing bacteria.

The equations for transport can then be written as follows (note that this is a three-dimensional form, thus the inclusion of the grad operator ∇):

$$\frac{\partial \theta C}{\partial t} = -\nabla \cdot \left(-D\theta\nabla C + \left(V_f + V_m + V_g\right)\theta C\right) + \left[\left(\frac{\mu_m C_f}{K_s + C_f}\right) - k_d\right] \cdot \theta C - \left[k_c(n-\sigma)C - k_y \rho\sigma^h\right] \tag{3}$$

$$\frac{\partial \theta \gamma_f}{\partial t} + \frac{\partial \rho s_f}{\partial t} = -\nabla \cdot \left(-D_f \theta \nabla C_f + \left(V_f \theta C_f\right)\right) - \frac{1}{Y}\left(\frac{\mu_m C_f}{K_s + C_f}\right)(\rho\sigma + \theta c) \tag{4}$$

$$s_f = k_a C_f^m \tag{5}$$

$$\frac{\partial \rho\sigma}{\partial t} = \left[\left(\frac{\mu_m C_f}{K_s + C_f}\right) - k_d\right]\rho\sigma + k_c(n-\sigma)C - k_y \rho\sigma^h \tag{6}$$

Parameters that must be specified to complete the formulation are

θ = volumetric moisture content
D = hydrodynamic dispersion coefficient
V_f = fluid velocity
V_m = chemotactic velocity of bacteria
V_g = settling velocity of bacteria
μ_m = maximum specific growth rate
K_s = half-saturation constant
k_d = death rate constant
k_c = pore-clogging rate constant
n = porosity
k_y = declogging rate constant
Y = yield coefficient
ρ = density of bacteria
h = filtration parameter
D_f = hydrodynamic dispersion coefficient for substrate
k_a = adsorption parameter
m = adsorption parameter

Taking the moisture content and water velocity as "known" or measurable quantities leaves 14 parameters to be estimated. As such, this model is not particularly useful in field situations. Furthermore, it does not take into account any sorption of contaminants to the particles (bacteria). Thus, it is a model for the transport of the particles, not of a contaminant undergoing facilitated transport.

This is still a reasonable start, because adequate models of particle transport have not been devised for all hydrogeological situations. Hornberger et al.[32] compared three basic modeling approaches to particle transport in which the governing equations of Dieulin (as discussed by de Marsily[30]), Rajagopalan and Chu,[31] and Corapcioglu and Haridas.[28] The Dieulin approach included dispersion of particles, but did not include any reaction term, i.e., no surface-surface interactions, retardation, filtering, etc. The approach of Rajagopalan and Chu[31] assumed that dispersion did not occur, but that reaction of the particles with the mineral grains did. Thus, these authors included a term for deposition of the particles on the medium grains, and they also included a term for the entrainment of particles back into the mobile fluid. Hornberger et al.[32] took the basic transport model of Corapcioglu and Haridas[28] and deleted all the terms that dealt with bacterial growth and decay to leave a straightforward transport model. The equations (using the same notation as equations 3 to 6 and written in a one-dimensional format) are

$$\frac{\partial C}{\partial t} = D\frac{\partial^2 C}{\partial Z^2} - V\frac{\partial C}{\partial Z} - k_c C + k_y s \tag{7}$$

$$\partial s / \partial = k_c C - k_y S \tag{8}$$

The results of the comparison of the approaches to data obtained in laboratory column studies indicated that any model of particle transport should include dispersion, and deposition/entrainment terms need to be included. For particles which are completely filtered by the medium, the entrainment term (k_y may be set to 0.

At this point, if it is assumed that particle transport can reasonably be modeled in the homogeneous case, the missing element is that of terms that adequately describe the competitive sorption/desorption reactions among the dissolved species and the various particle surfaces, mobile and immobile, with which the contaminant could associate.

Mills et al.[41] published a model (COMET) which added to the particle-transport equations a series of equations to describe sorption of metals to surfaces in the aquifer. This model assumed that sorption was an equilibrium process; however, and in the case of moving water, the assumption of equilibrium may be often violated. A recent formulation (referred to as COMATOSE) has been prepared[43] which treats the surface-surface interactions as kinetic processes. This approach is more satisfying in terms of the probable mechanisms that operate to govern facilitated transport.

VII. CONCLUSION

It should be clear that transport of contaminants as particle-associated forms or as pure colloids is still an open issue. The potential for significant movement of materials is apparent, yet the actual importance is only partially known. Theoretical models are now being developed that can describe such transport, but the modeling effort is hampered by a lack of information on the mechanisms of particle formation, particle-contaminant association, and particle transport. While some very good work has been done, many questions remain. Design and management of effective waste repositories for radionuclides, heavy metals, and toxic organics demands adequate consideration of facilitated transport to ensure successful containment of the wastes contained therein.

REFERENCES

1. Stumm, W. and Morgan, J. J., *Aquatic Chemistry*, John Wiley and Sons, New York, 1981.
2. Borchardt, J. A. and O'Melia, C. R., Sand filtration of algal suspensions, *J. Am. Water Works Assoc.*, 80, 36, 1961.
3. Ives, K. J., Filtration using radioactive algae, *Proc. Am Soc. Civ. Eng.*, 87, 23, 1961.
4. Eliassen, R., Clogging of rapid filters, *J. Am. Water Works Assoc.*, 33, 926, 1941.
5. Hunter, R. J. and Alexander, A. E., Surface properties and flow behavior of kaolinite. Electrophoretic mobility and stability of kaolinite sols, *J. Colloid Interface Sci.*, 18, 820, 1963.
6. Hunter, R. J. and Alexander, A. E., Surface properties and flow behavior of kaolinite. Electrophoretic studies of anion adsorption, *J. Colloid Interface Sci.*, 18, 833, 1963.
7. Hunter, R. J. and Alexander, A. E., Surface properties and flow behavior of kaolinite. Flow of kaolinite sols through a silica column, *J. Colloid Interface Sci.*, 18, 842, 1963.
8. Matijevik, E., Kuo, R. J., and Kolny, H., Stability and deposition phenomena of monodispersed hematite sols, *J. Colloid. Interface Sci.*, 80, 94, 1981.
9. Ives, K. L. and Gregory, J., Surface forces in filtration, *Proc. Soc. for Water Treat. Examination*, 15, 93, 1966.
10. Fitzpatrick, J. A. and Spielman, L. A., Filtration of aqueous latex suspensions through beds of glass spheres, *J. Colloid Interface Sci.*, 43, 350, 1973.
11. Kuo, R. J. and Matijevik, E., Particle adhesion and removal in model systems. III. Monodispersed ferric oxide on steel, *J. Colloid. Interface Sci.*, 78, 407, 1980.
12. Kallay, N. and Matijevik, E., Particle adhesion and removal in model systems. IV. Kinetics of detachment of hematite particles from steel, *J. Colloid Interface Sci.*, 83, 289, 1981.
13. Champlin, J. B. F. and Eichholz, G. G., The movement of radioactive sodium and ruthenium through a simulated aquifer, *Water Resour. Res.*, 4, 147, 1986.
14. Eichholz, G. G., Wahlig, B. G., Powell, G. F., and Craft, T. F., Subsurface migration of radioactive waste materials by particulate transport, *Nuclear Technol.*, 58, 511, 1982.

15. Champp, D. R., Meritt, W. F., and Young, J. L., Potential for rapid transport of Pu in ground water as demonstrated by core column studies, in *Scientific Basis for Radioactive Waste Management*, Elsevier, New York, 5, 745, 1982.
16. Torok, J., Buckley, L. P., and Woods, B. L., The separation of radionuclide migration by solution and particle transport in soil. *J. Contam. Hydrol.*, 6, 185, 1990.
17. Mahara, Y. and Matsuzuru, H., Mobile and immobile plutonium in a groundwater environment, *Water Resour. Res.*, 23, 43, 1989.
18. Giblin, A. M., Batts, B. D., and Swaine, D. J., Laboratory simulation studies of uranium mobility in natural waters, *Geochim. Cosmochim. Acta*, 45, 699, 1981.
19. Bates, J. K., Bradley, J. P., Teetsov, A., Bradley, C. R., and Buchholtz ten Brink, M., Colloid formation during waste form reaction: implications for nuclear waste disposal, *Science*, 256, 649, 1992.
20. Buddemeier, R. W. and Hunt, J. R., Transport of colloidal contaminants in groundwater: radionuclide migration at the Nevada Test Site, *Appl. Geochem.*, 3, 535, 1988.
21. Penrose, W. R., Polzer, W. L., Essington, E. H., Nelson, D. M., and Orlandini, K. M., Mobility of plutonium and americium through a shallow aquifer in a semiarid region, *Environ. Sci. Technol.*, 24, 228, 1990.
22. Magaritz, M., Amiel, A. J., Ronen, D., and Wells, M. C., Distribution of metals in a polluted aquifer: a comparison of aquifer suspended material to fine sediments of the adjacent environment, *J. Contam. Hydrol.*, 5, 333, 1990.
23. Magee, B. R., Lion, L. W., and Lemley, A. T., Transport of dissolved organic macromolecules and their effect on the transport of phenanthrene in porous media, *Environ. Sci. Technol.*, 25, 323, 1991.
24. Enfield, C. G., Bengtsson, G., and Lindqvist, R., Influence of macromolecules on chemical transport, *Environ. Sci. Technol.*, 23, 1286, 1989.
25. Brownawell, B. J. and Farrington, J. W., Biogeochemistry of PCBs in interstitial waters of a coastal marine sediment, *Geochim. Cosmochim. Acta*, 50, 157, 1986.
26. Dunnivant, F. M., Jardine, P. M., Taylor, D. L., and McCarthy, J. F., Cotransport of cadmium and hexachlorobiphenyl by dissolved organic carbon through columns containing aquifer material, *Environ. Sci. Technol.*, 26, 360, 1992.
27. Corapcioglu, Y. M. and Haridas, A., Transport and fate of microorganisms in porous media: a theoretical investigation, *J. Hydrol.*, 72, 149, 1984.
28. Corapcioglu, Y. M., and Haridas, A., Microbial transport in soils and groundwater: a numerical model, *Adv. Water Resour.*, 8, 188, 1985.
29. McDowell-Boyer, L. M., Hunt, J. R., and Sitar N., Particle transport through porous media, *Water Resour. Res.*,22, 1901, 1986.
30. de Marsily, G., *Quantitative Hydrogeology*, Academic Press, San Diego, 1986.
31. Rajagopalan, R. and Chu, R. Q., Dynamics of adsorption of colloidal particles in packed beds, *J. Colloid Interface Sci.*, 86, 299, 1982.
32. Hornberger, G. M., Mills, A. L., and Herman, J. S., Bacterial transport in porous media: evaluation of a model using laboratory observations, *Water Resour. Res.*, 3, 915, 1992.
33. Yates, M. V. and Yates, S. R., Modeling microbial transport in the subsurface: a mathematical discussion, in *Modeling the Environmental Fate of Microorganisms*, Hurst, C. J., Ed., Am. Soc. Microbiol., Washington, D.C., 1991, 48.
34. Freeze, R. A. and Cherry, J. A., *Groundwater*, Prentice-Hall, Englewood Cliffs, NJ, 1979.

35. Harvey, R. W. and George, L. H., Transport of microspheres and indigenous bacteria through a sandy aquifer: results of natural- and forced-gradient tracer experiments, *Environ. Sci. Technol.*, 23, 51, 1989.
36. Puls, R. W. and Powell, R. M., Transport of inorganic colloids through natural aquifer material: implications for contaminant transport, *Environ. Sci. Technol.*, 26, 614, 1992.
37. Saiers, J. S., The Influence of Physical and Chemical Factors on the Transport and Fate of Inorganic Colloids in Porous Media, M.S. thesis, University of Virginia, Charlottesville, 1991.
38. Elimelech, M. and O'Melia, C. R., Kinetics of deposition of colloidal particles in porous media, *Environ. Sci. Technol.*, 24, 1528, 1990.
39. Germann, P. F., Smith, M. S., and Thomas, G. W., Kinematic wave approximation to the transport of *Escherichia coli* in the vadose zone, *Water Resour. Res.*, 23, 1281, 1987.
40. Mints, D. M., Filtration kinetics of a low concentration suspension in water of water clarification filters, *Dokl. Akad. Nauk. SSSR*, 78, 315,1951.
41. Mills, W. B., Liu, S., and Fong, F. K., Literature review and model (COMET) for colloids/metals transport in porous media, *Groundwater*, 29, 199, 1991.
42. Wang, H. F. and Anderson, M. P., *Introduction to Groundwater Modeling: Finite Difference and Finite Element Methods*, W. H. Freeman and Co., San Francisco, 1982.
43. Saiers, J. S. and Hornberger, G. M., in preparation.

CHAPTER 5

Trace Metal-Suspended Particulate Matter Associations in a Fluvial System: Physical and Chemical Influences

Lesley A. Warren and Ann P. Zimmerman

TABLE OF CONTENTS

0-87371-678-7/93/$0.00+$.50

I. OVERVIEW OF TRACE METAL-PARTICULATE INTERACTIONS

A. The Importance of Suspended Particulates

It has been widely recognized that the fate and transport of metals within aquatic systems is determined by the nature of their association with suspended particulate matter (SPM).[1-9] SPM, in most systems, represents some combination of inorganic material, such as clays, hydrous metal oxides (principally Fe and Mn oxides), and organic matter, both detrital and living. Because of the large surface areas of silts, clays and organic matter available for sorption, and the scavenging nature of oxides,[10-12] SPM is often the major transport vector for heavy metals.[13-15] While particulate-associated trace-metal concentrations tend to be substantially higher than those in solution, SPM concentrations are substantially lower relative to the volume of water in fluvial systems, so solution complexes tend to dominate the total metal load.[16]

B. The Mechanism of Trace Metal-Particulate Association

1. Sorption Reactions

Surface sorption reactions involving trace metals at the solid-solution interface are thought to be the principal mode of metal particulate association in natural aquatic systems.[3,9,11,15,17,18] Metals may be "sorbed" onto inorganic surfaces, such as oxides or clays, or complexed with inorganic or organic ligands. There are no clear analytical boundaries between the various types of sorption reactions. Furthermore, the mechanisms for different sorption processes as well as for precipitation/coprecipitation can rarely be unambiguously characterized.[17,19,20]

Idealized sorption mechanisms are distinguished as: (1) physical adsorption — due to nonspecific forces of attraction (van der Waals forces) involving the entire electron shells of the dissolved trace element and of the adsorbent; (2) electrostatic adsorption — (ion exchange) due to coulombic forces of attraction between charged solute species and the adsorbing phase; (3) specific adsorption — (chemisorption) due to the action of chemical forces of attraction leading to surface bonds to specific sites on the solid phase; and (4) chemical substitution — (coprecipitation or solid solution) can be compared with adsorption reactions in the sense that the result of these processes would be a removal of a trace constituent from the solution phase.[19,21,22]

2. Factors Influencing Sorption Reactions

Sorption processes are influenced by physical and chemical parameters, such as pH, pE (redox), temperature, types and concentrations of complexing agents, trace-element concentration and speciation, binding affinities, ionic strength, and the composition and surface properties (e.g., charge) of the solid phase. pH is widely recognized as the master variable governing the extent of sorption, as both particulate surface acidity and hydrolysis of metal ions are pH dependent.[19,23-26]

C. Oxides

1. Importance of Oxides

Oxides in particular have received a great deal of attention in the literature due largely to their abundance, ubiquitous nature, and ion-exchange properties.[22,27-30] Metals, such as iron, aluminum, and manganese, readily precipitate in natural waters to form particulate or colloidal hydroxides (see Figure 1.; note: oxides, hydroxides, and oxyhydroxides are probably synonymous terms from an analytical perspective).

2. Oxide Trace Metal Complexation

Oxide sorption of metal ions is understood as a competitive complex formation involving one or two surface hydroxyls (Figure 2). A number of surface-metal complexes are possible: (1) a metal ion can be complexed by the O-donor atom of one or two surface hydroxyls; (2) an O-donor atom can complex a metal-ligand complex; (3) the underlying metal ion (e.g., Fe, Mn, Si) can exchange its surface hydroxyl group for another ligand; or (4) the underlying metal ion can exchange its OH group for a ligand-metal complex (Figure 2).

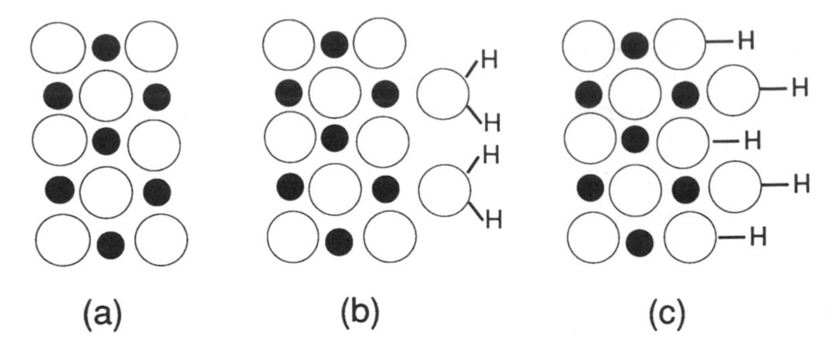

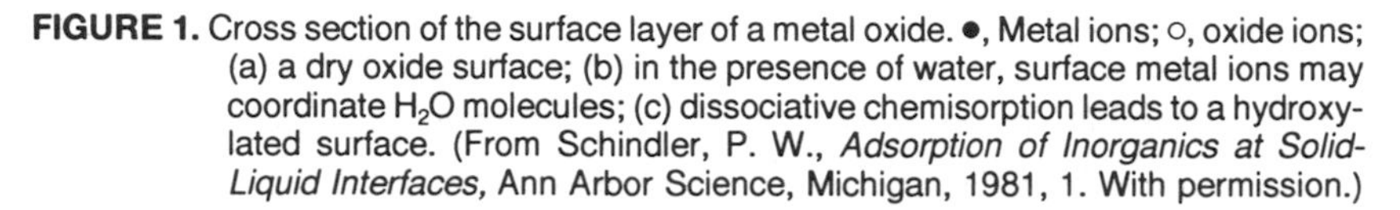

FIGURE 1. Cross section of the surface layer of a metal oxide. •, Metal ions; ○, oxide ions; (a) a dry oxide surface; (b) in the presence of water, surface metal ions may coordinate H_2O molecules; (c) dissociative chemisorption leads to a hydroxylated surface. (From Schindler, P. W., *Adsorption of Inorganics at Solid-Liquid Interfaces,* Ann Arbor Science, Michigan, 1981, 1. With permission.)

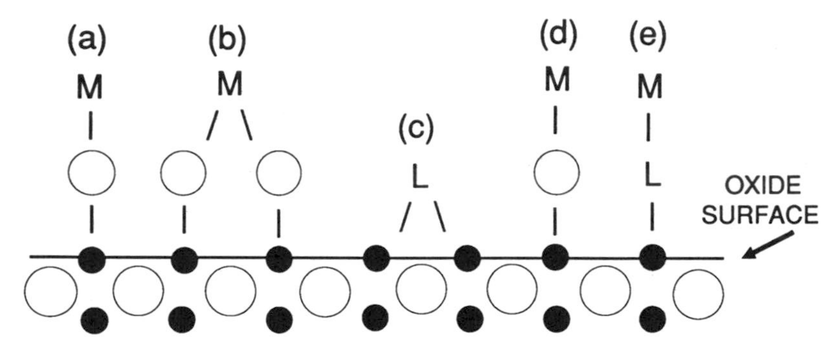

FIGURE 2. Possible hydrated oxide surface-complexation reactions. Oxide surface: •, metal ions; ○, oxide ions. Solution ions: M, metal ion; L, ligand; (a) the surface hydroxyl group (OH) has a complex forming O-donor atom that can complex a metal ion from solution; (b), two surface O-donor atoms can complex one metal ion; (d), or the O-donor atom can complex a metal-ligand complex; (c), the underlying metal ion (•, e.g., Fe, Mn, Al, Si), can exchange its surface OH group for other ligands; (e), or the underlying metal ion can exchange its OH group for a ligand-metal complex.[35, 71, 107]

Surface oxide sites can be considered analogous to solution ligands; however, their double layer (a function of pH and ionic strength) also influences sorption. Oxides are amphoteric Bronsted acids; therefore, in a basic environment, oxide surfaces are cation exchangers as the surface charge is negative. Thus, in high pE environments, oxides are capable of sorbing significant quantities of trace metals[25,31-38]

D. Particulate Organic Matter (POM)

1. Particle Organic Coatings

The role of natural particulate organic matter (POM) in trace metal adsorption is not well understood.[39] The large size and low solubility of humic compounds suggest that these organic compounds have an affinity for the particulate surface-water interface, and generally it is accepted that most particles in aquatic systems are coated with a film of organic matter,[40-42] of which the major components are humic-like refractory organic substances (ROS).[43] The adsorbed organic material is at least partly responsible for observations that the majority of particles are negatively charged when suspended in water.[44] ROS adsorption by oxides is thought to occur by ligand exchange with H_2O or OH^- groups and/or by surface complexation between oxide OH_2 groups and ROS-anionic groups.[43,44]

2. *Oxide-POM Interactions*

Studies provide evidence supporting both competition between oxide and organic sediment pools for trace metals, as well as organic enhancement of oxide metal sorption.[44-49] Both oxides and organic matter are known to be efficient scavengers of trace metals. Since equilibrium calculations generally demonstrate particulate binding sites to be in excess in most aquatic systems, competition for trace metals between these two sediment pools could occur if all binding sites have an equal affinity for each metal. The existence of two discrete sedimentary pools — oxides and organics — is in itself an operational artifact. If the majority of the organic pool is itself sorbed to oxide surfaces, its presence could either create new sorption sites or stabilize previously nonfunctioning sites on the oxide. In either case, the sorptive capacity of the oxide is enhanced. ROS may also serve as a transport vector for trace metals from the colloid/solution phases to the particulate oxide surface.

E. Comparison of Trace Metal Complexation by POM and Oxides

In general, observations of metal-binding characteristics of oxides and organic compounds have revealed a number of similarities: (1) binding intensity varies as a function of the degree of site occupation (e.g., decreased equilibrium constants at greater adsorption densities, suggesting distinct groups of binding sites with differing binding affinities); (2) binding intensity depends on the cation, suggesting strong binding sites for one metal are not necessarily preferred binding sites for other metal ions; and (3) binding energy sequences for specific cations are ligand specific, varying as a function of the ligand system.[8,25,34,36,50-54]

F. Trace Metal Speciation

Speciation of trace metals is of great ecotoxicological interest, since the uptake and toxicity of metals are often better correlated with the concentration of specific metal species than with the total metal concentration. In many cases, the bioavailability and toxicity of the free-hydrated ion is greater than that of metals complexed with other ligands.[55-58] In addition, the form in which metals enter the sediment phase is important for determining the rate and extent of redistribution and/or release.[15,59] Solubility, mobility, and potential bioavailability of particle-bound metals can be increased by four major factors: (1) lowering of pH below the pH adsorption edge (a narrow range of approximately 1 to 2 pH units where adsorption shifts between 0 and 100%);[20,29] (2) an increased occurrence of natural and/or synthetic complexing agents (e.g., EDTA);[50] (3) increased salt concentration (competition with metals for sorption sites and formation of soluble chlorocomplexes with some trace metals can reduce the formation of particulate metal complexes);[60] (4) changing redox conditions (such changes can affect metals directly by changing the oxidative state of the metal ion or indirectly by changing the availability of binding sites on competing ligands or chelates).[33]

G. The Influence of Particle Grain Size

Sorption reactions are influenced by physical sediment characteristics such as grain size. Various geochemical components of the suspended sediment fraction often occur in characteristic size ranges with implications for differential transport, deposition, sorption, and/or rerelease of heavy metals.[61-63] POM and oxides tend to concentrate in the smaller size fractions.[16,46,47] Finer grained sediments adsorb relatively more contaminants per unit volume or mass due to their relatively larger surface-to-mass ratios, and hence exhibit relatively larger adsorptive capacities.[64-66]

H. Trace Metal-SPM Interactions in Natural Systems

Most of our current knowledge about aqueous surface chemistry has come primarily from studies on equilibrium metal adsorption in model laboratory systems using well-characterized model particles (predominantly Fe oxides, e.g., Dzombak and Morel.[67] Attempts at modeling surface complexation in natural environments are hampered by their inherent complexity and dynamic nature. Natural systems are heterogeneous environments that are both spatially and temporally variable, presenting a number of sampling and analytical difficulties. Yet, field studies are a critical component of models for natural systems, especially field studies that contribute to the theoretical framework for trace metal particulate reactions in natural environments.

II. METHODOLOGY

A. SPM Collection

Collection of an SPM sample is critical to subsequent interpretation of associated trace metal concentrations, geochemical partitioning, and transport. The aim of a given study will define the sampling protocol; however, all aim to achieve a representative sample. The reader is referred to Horowitz[68] for an excellent review. The discussion here will be confined to the <63-μm fraction of the suspended sediment pool, as this size class is considered to be the important fraction for geochemical studies.[68] Particles less than 63 μm are believed to be evenly distributed in fluvial cross sections, and thus samples collected near the centroid of flow will provide a representative sample for any given time.[16] If the ultimate aim of a study is to determine metal loads (i.e., mass transported), then it is critical to also measure discharge in the system.

A variety of methods can be used to collect SPM material (see Horowitz[68] for a complete discussion). The focus here is on continuous flow centrifugation, which can be done *in situ* or in the laboratory, once a water sample has been collected. Caution should be exercised when collecting *in situ,* as flow rate into the centrifuge will determine the size fractions collected. Without *a priori* knowledge of the size distribution, it is prudent to select a relatively low flow rate concomitant with a high RPM factor to ensure collection of the finer grained particles. The advantages to collection in the field are speed of collection and no water transportation difficulties. Collection of SPM in the laboratory requires that a water sample be collected in the field, brought back to a laboratory, and spun there. If SPM concentrations are low, this can result in the collection of thousands of liters of water. Whichever method is chosen, these methods of collection result in an "instantaneous" sample. Temporal variability, (e.g., Warren and Zimmerman[69]) can only be addressed by collecting samples over a long period of time (i.e., >8 h[68]) or by increasing the frequency at which samples are collected. No technique or collector will cover both spatial and temporal variability. Again, the approach will be determined by the question being addressed.

B. Physical Characteristics of SPM

1. Particle Size Distribution

SPM physical characteristics are highly influential in trace metal sorption reactions. Particle size distribution and surface areas associated with SPM are important variables to consider. Particle size distribution can be determined by several methods, e.g., physical separation by sieving techniques, pipet analysis, centrifugation, or direct measurement of the grain sizes present in a sample by microscopy, image analyzers, or sedigraphs.[68] Thus, particle size distributions are operationally defined by the method employed, and results will be dependent upon which technique is used. Direct measurement techniques are faster and more information rich, yielding a continuous distribution of grain size. However, analytical protocols involving specific size classes will require use of a physical separation technique.

2. SPM Surface Area

Given that sorption reactions at particle surfaces play a key role in trace-metal complexation, associated surface area of SPM is one of the more important variables to measure. In general, field studies have expressed metal concentrations normalized by weight, i.e., μg metal per gram of sedimentary material. If surface reactions control particulate trace metal reactions, then it is not the mass of the sediment component that is important; it is the associated surface area that influences the concentrations of associated trace metals.

Several techniques exist to measure surface areas, and results must be considered operationally defined. The majority of techniques involve gas adsorption (e.g., BET-nitrogen/helium).[70-72] However, surface area can also be determined using a laser particle-sizing system. This approach has several advantages: (1) water samples are untouched before analysis (thus maintaining the SPM characteristics *in situ*); (2) concentrated samples are not required; (3) measurement is relatively quick, which coupled with the previous point, facilitates replication; and (4) distributions can be calculated as a function of particle number, surface area, or volume.

C. Geochemical Fractionation

1. Extraction Schemes

Evaluating the role of discrete sediment phases (e.g., oxides, organics, carbonates) in determining the ultimate fate of trace metals requires partitioning of trace metals among those phases. However, it is not currently possible to physically separate discrete geochemical entities in a heterogeneous, natural sediment. Thus, the concentrations of trace metals associated with given geochemical phases are most commonly determined using operational extraction/fractionation schemes. There is a considerable body of evidence supporting the utility of extraction schemes.[66,73-78] The goal of all such schemes is to precisely determine the partitioning of metals among discrete sediment phases considered important in controlling metal concentrations, e.g., oxides, organic matter, carbonates. The main criticism of such schemes is the possibility of post-extraction resorption[79,80] (i.e., cation extraction into one reagent followed by resorption onto another sediment phase and subsequent reextraction from that phase). However, post-extraction resorption of Cd, As, Ni, Cu, and Zn in real sediments was found to be negligible.[81] Additional metals should be similarly investigated, but the lack of an operational effect for two metals as dissimilar as Cd and Cu in their sorption preferences (oxides vs. organic matter), tends to support the efficacy of the extraction approach.

2. Phases of Interest in Extraction Schemes

Probably the most commonly used extraction scheme (or modification of) is that of Tessier et al.[78] The Tessier protocol extracts metals associated with five operationally defined phases: (1) "exchangeable", e.g., adsorbed at particle surfaces; (2) "carbonate associated", e.g., discrete carbonate minerals or coprecipitated with major carbonate phases; (3) "oxides", e.g., occluded in Fe and/or Mn oxyhydroxides, either as discrete nodules or as coatings on particles (this step can be done in two stages to separate metals associated with Mn oxides from those associated with Fe oxides); (4) "organics and sulphides", e.g., bound up with organic matter, in either living or detrital form, or bound in amorphous authigenic sulfides or in more crystalline forms; (5) "bound within mineral lattices", e.g., bound in lattice positions in aluminosilicates, in resistant oxides, or in resistant sulfides.[82] These fractions are usually chosen as the literature suggests that the possible interactions responsible for metal control are (1) metal-clay colloid ion exchange reactions, (2) metal sorption onto hydrous oxides, (3) metal sorption onto humic acid-clay colloids, (4) metal-inorganic anion complexes, and (5) metal-organic ligand complexes.[83]

3. Operational Definition of Extracted Metals

A critical assumption for extraction schemes is that the reagents used are selective, e.g., the solvent is either phase or mechanistically specific. Results for any scheme will be operationally defined by the extractants used. Therefore, it is more appropriate to describe metals extracted using such schemes by the reagents used, rather than to ascribe metals to a true geochemical fraction that the scheme is supposed to represent. Therefore, Tessier's "exchangeable and carbonates associated metals" are best described as leachable metals; "oxide associated metals" are best described as reducible metals; "organics and sulphides associated metals" are literally oxidizable metals; and "resistate mineral associated metals" are residual metals (defined by the reagent and conditions).

4. Sequential-Simultaneous Extraction Schemes

The vast majority of results generated using extraction schemes follow a sequential approach, where the same aliquot of sediment material is subjected to sequentially applied, stronger extractants. This approach is useful where sediment concentrations are low. However, it requires a minimum of 3 to 4 d to complete the extraction scheme. A newer approach, e.g., Bendell-Young and Harvey,[84] is to use a simultaneous extraction scheme where separate aliquots are used for each extraction step, and the total for each fraction is determined by difference. Each extractant represents the amount of metal present in that particular phase, plus any of the more loosely associated phases. Therefore, exchangeable metal concentrations should contain the same amount of total metal, whether sequentially or simultaneously extracted, while a simultaneously extracted oxidizable sample contains the total amount of metal from the leachable, reducible, and true oxidizable metal fractions. Consequently, determining the actual amount of metal in a given phase in a simultaneous extraction requires subtraction of the previous phase's concentration (e.g., reducible metal = reducible total minus the leachable component; oxidizable metal = oxidizable total metal extracted minus the leachable and reducible components; and residual metal = residual total metal extracted minus the other three phases). This approach requires more material, but all steps can be carried out simultaneously, thus reducing extraction time to 1 d, and the possibility of redistribution over time.

5. *Reference Materials*

There are no standard reference materials for extraction procedures; nevertheless, NBS river sediment reference materials (National Institute of Standards and Technology, U.S. Department of Commerce) can be used in the extraction scheme to check for percent recovery of metals. In this case, the NBS material is run through the extraction scheme and the totals tabulated from each phase to check against the total certified value. Use of standard reference materials is highly recommended.

III. A CASE STUDY: THE DON RIVER

A. Don River Basin

The Don River (LAT 79° 20′, LONG 43° 55′) is a relatively small system approximately 32 km long, draining a highly urbanized watershed of 376 km^2. It is one of numerous small rivers draining the Metropolitan Toronto area, emptying into the Toronto Harbour of Lake Ontario. Increasing contamination of the nearshore Lake Ontario waters (an International Joint Commission Area of Concern) indicates a serious metal-transport problem into Lake Ontario by such rivers. There are over 1400 street run-off and discharge outfalls on the Don, as well as a sewage treatment plant and a snowmelt yard. A major traffic corridor runs along the river for approximately one third of its length. The Don Basin is composed chiefly of Wisconsin till and subsequent deposits, resting either directly on bedrock (limestone and shales), or else on unconsolidated material deposits from previous ice invasions. In the upper reaches of the watershed, the Don is split into two branches, the West Don and the East Don, these two branches converge approximately 8 km from the mouth of the river into the Main Don (Figure 3). The West Don Basin has a greater proportion of coarser deposits — sand and gravel, in comparison with the East Don Basin which has large regions of clay and silt. The Main Don dissects the shallow Lake Iroquois bed and also has proportionally higher amounts of clay and silt than does the West branch. Land-use patterns shift from predominantly agricultural in the headwater regions of the West Don to predominantly residential, with some commercial use lower down on the West Don. The East and Main Don branches show similar land-use patterns of mixed residential/industrial/commercial use, though the latter section of the river also receives inputs from the sewage treatment plant and the snowmelt yard.[85-87] The Don is a hardwater, high alkalinity system, with high pH (approximately 8) and oxidizing pE.

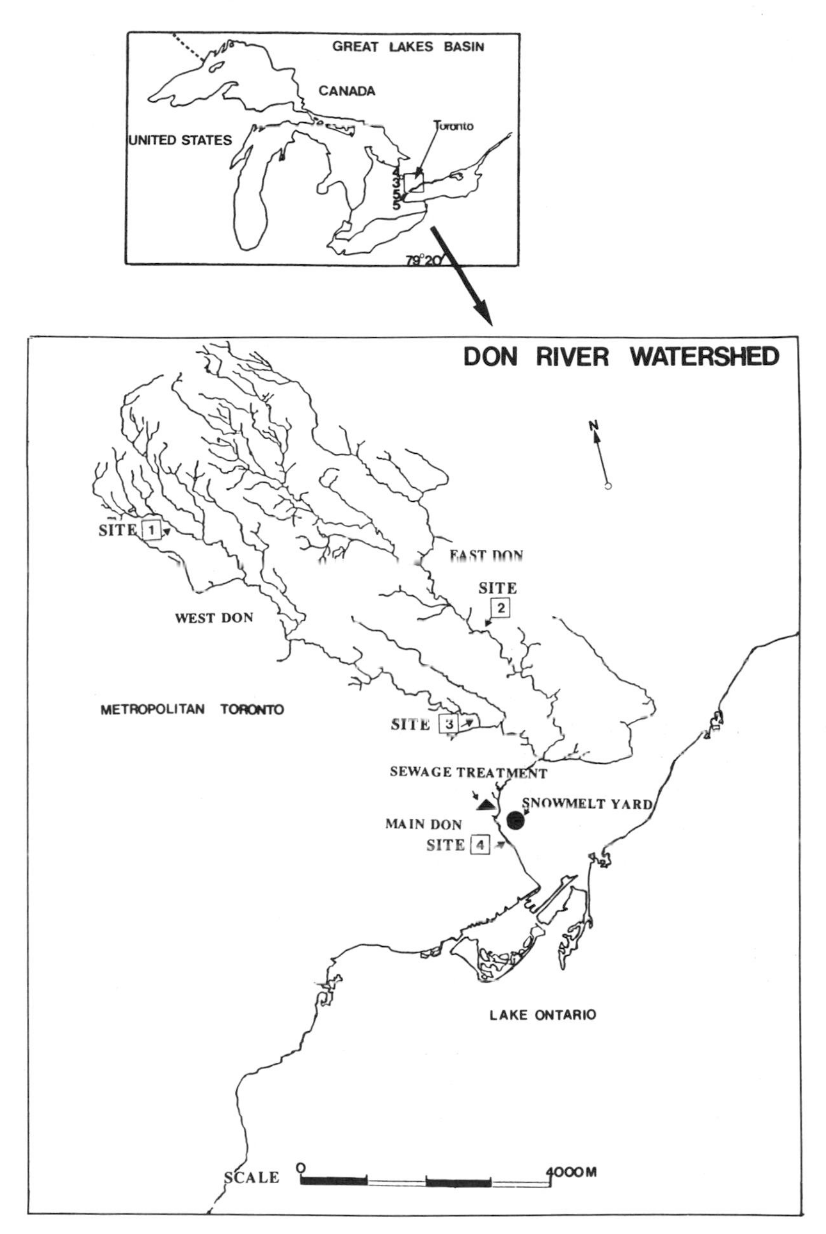

FIGURE 3. The Don River Basin, showing location of sampling sites, sewage treatment plant, and snowmelt yard.

B. Sample Collection

A total of 37 river-water samples (approximately 200 to 400 l) over the period from March 1988 to April 1989 were collected at four sites (Figure 3) chosen to maximize the differences in basin characteristics. Subsequent collection of SPM was achieved by continuous flow centrifugation in the laboratory. SPM samples were isolated within 24 h of collection and then run through an extraction scheme modified from Tessier et al.[78] to extract metals in the following four fractions: (1) leachable ("exchangeable and carbonates associated"); (2) reducible ("oxides associated"); (3) oxidizable ("organics associated" — note: Don River SPM exists in an oxic environment; sulphides are not considered an important phase); and (4) residual ("bound within mineral lattices").

C. Patterns of Trace Metal Geochemical Partitioning

As expected, absolute concentrations of metals vary across sites and samples. Nevertheless, patterns of geochemical partitioning are spatially and temporally stable (Figures 4 to 7). The highest concentrations of Cd are associated with the leachable and reducible phases across all sites (Figure 4). Cd is the only metal which shows a substantial proportion consistently in the leachable phase. Highest Cu concentrations are associated with the oxidizable and residual phases across samples and sites (Figure 5). Cu is the only metal with a consistently high association with the oxidizable fraction. Zn is consistently characterized by a strong association with the reducible phase (Figure 6). Some oxidizable-associated Zn occurs periodically, depending on site and sample. Negligible amounts of leachable-associated Zn were found. Pb also shows a strong affinity for the reducible phase (Figure 7). Virtually no Pb is detected associated with the leachable phase, but consistently high proportions are found associated with the residual phase.

Since Cd is more weakly complexed and thus more readily desorbed, it is expected to be most sensitive to changes in system chemistry. Cu, Pb, and Zn are more strongly complexed than is Cd in the Don. However, both Pb and Zn, in addition to Cd, show a strong affinity for the reducible phase ("oxides" associated) which can be extremely sensitive to changes in pE (i.e., lowered pE may cause reduction of oxides and subsequent release of their associated metal loads; such changes may occur once the Don River enters the Toronto Harbor and particulates settle into anaerobic environments).

Three patterns of trace-metal affinity for particulate geochemical phases emerge for the four cations: (1) Cd — reducible > leachable > oxidizable; (2) Cu — oxidizable > leachable > reducible; and (3) Zn, Pb — reducible > oxidizable > leachable. These results from the Don River show similar patterns to those found by other authors for other systems.[88-91] Thus, the affinities of various geochemical phases for specific cations appear consistent across systems, suggesting an order to trace metal particulate complexation amenable to modeling and prediction.

D. Scavenging Abilities of POM and Oxides

The overall importance of a particular geochemical fraction of the SPM pool is related to its abundance relative to other sedimentary constituents.[92,93]

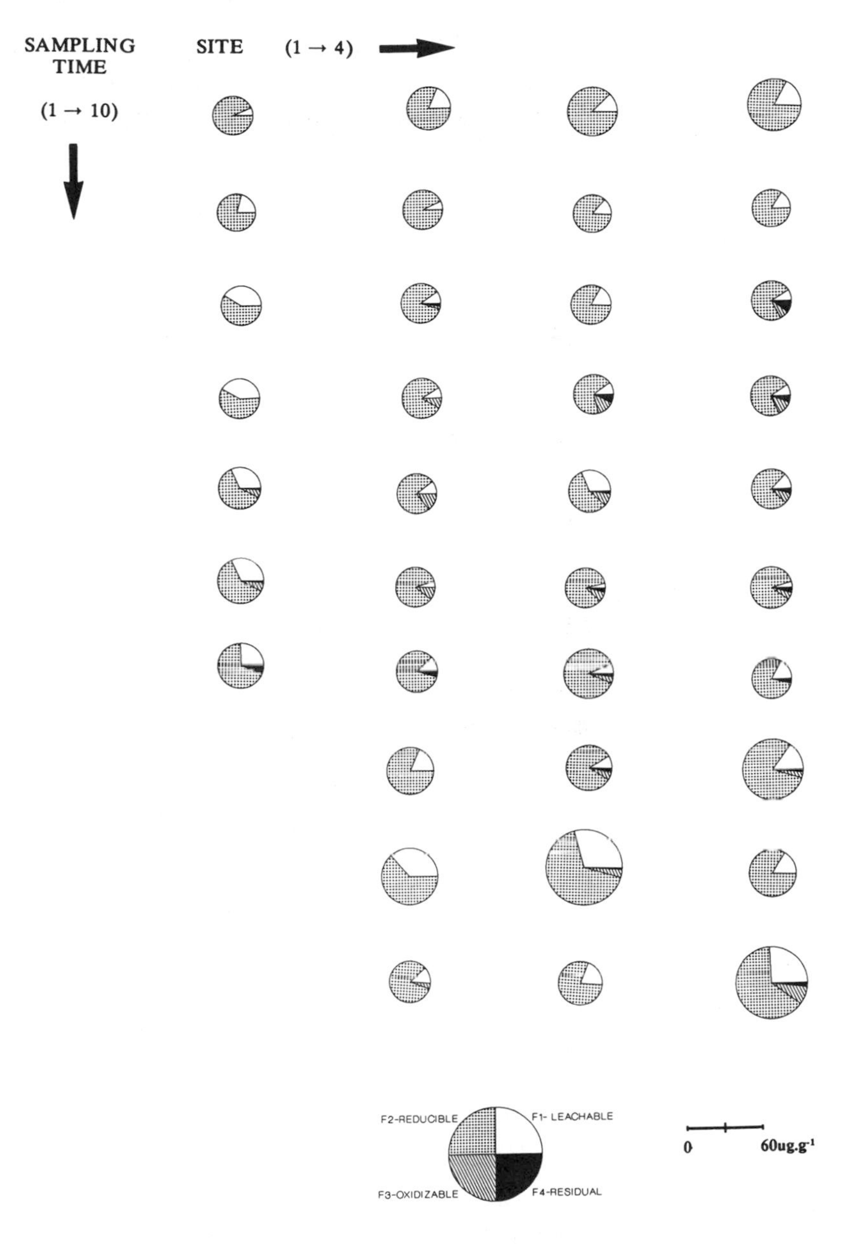

FIGURE 4. Proportions of total Cd found in four geochemical phases across samples at each of four sites. The diameter of each pie is proportional to the total concentration (note scale).

However, the scavenging ability of a particular sediment component, which is likely a function of associated surface area, may far exceed its relative contribution to the mass of SPM material. Comparing the scavenging abilities of discrete particulate geochemical phases under the fluctuating chemical matrix in the fluvial environment requires that the relative sorptive capacity be

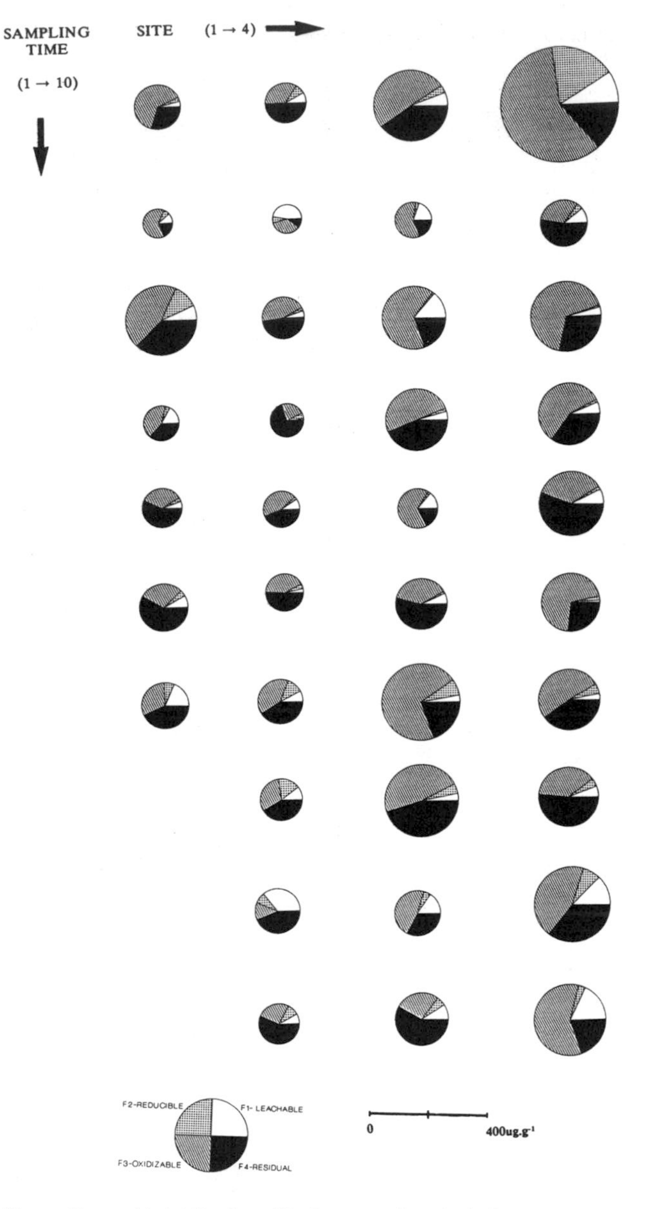

FIGURE 5. Proportions of total Cu found in four geochemical phases across samples at each of four sites. The diameter of each pie is proportional to the total concentration (note scale).

assessed, irrespective of absolute concentrations. Phase Concentration Factors (PCF; metal bound to particulate phase/particulate phase[94]) are a useful tool for this purpose in field studies, as surface areas of discrete sedimentary components can not yet be determined. PCFs for each metal (Cd, Cu, Zn, and Pb) were determined for both the reducible (PCF-r) and oxidizable (PCF-o) suspended-

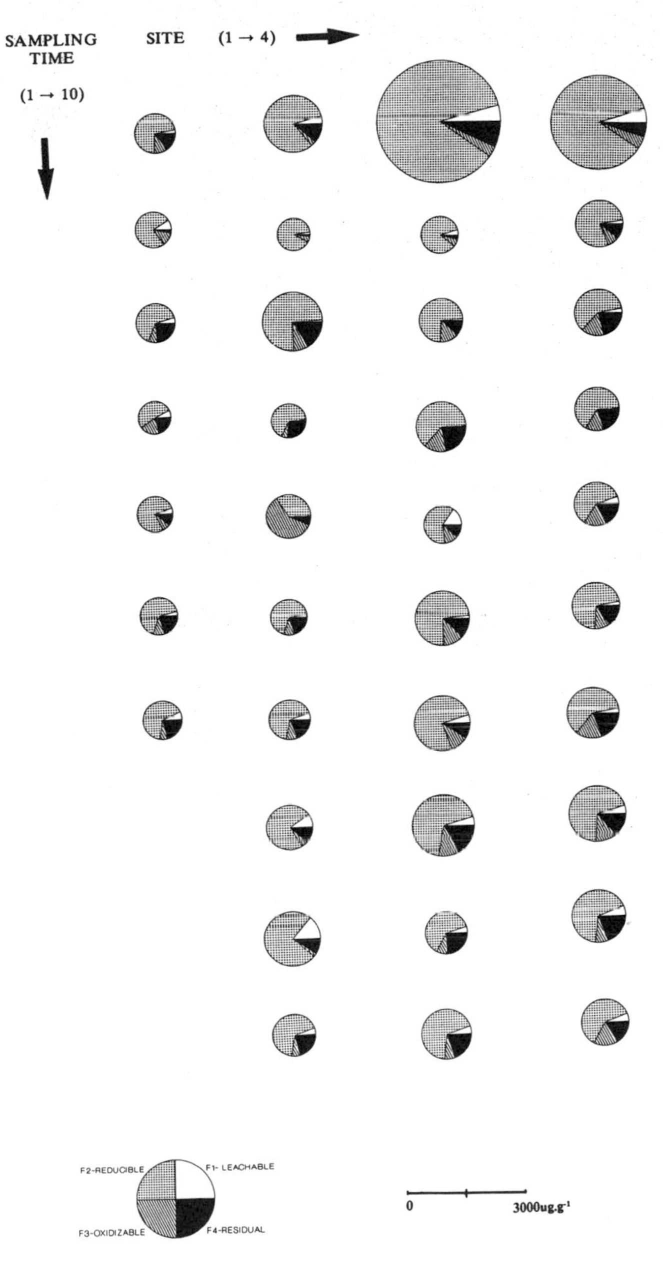

FIGURE 6. Proportions of total Zn found in four geochemical phases across samples at each of four sites. The diameter of each pie is proportional to the total concentration (note scale).

particulate fractions for the dataset described above. Cd, Zn, and Pb values for PCF-r were significantly higher than their respective PCF-o values at all four sites (Table 1). No significant differences, at any site, between PCF-r and

FIGURE 7. Proportions of total Pb found in four geochemical phases across samples at each of four sites. The diameter of each pie is proportional to the total concentration (note scale).

PCF-o for Cu emerged (Table 1). The SPM-oxide pool outcompetes the SPM-organic pool for Cd, Zn, and Pb, but sorbs Cu with the same relative concentration factor as does POM. Stability sequences for metal organic ligands

Table 1: Summary Statistics from General Linear Models (GLM) Analyses Comparing Mean PCF-r and PCF-o Values across Sites for Cd, Cu, Zn, and Pb

	Site	F	p	df	r2	PCF-r	PCF-o
Cd							
	1	24.74	0.0003	13	0.67	0.268	0.0004
	2	59.26	0.0001	19	0.77	0.289	0.0005
	3	2.61	0.12	19	0.13	0.781	0.006
	4	7.46	0.01	19	0.29	0.786	0.004
Cu							
	1	NS	—	—	—	0.324	0.166
	2	NS	—	—	—	0.286	0.137
	3	NS	—	—	—	0.294	0.783
	4	NS	—	—	—	0.534	0.632
Zn							
	1	26.59	0.0002	13	0.69	15.159	0.128
	2	11.03	0.003	19	0.38	28.945	0.316
	3	33.58	0.0001	19	0.65	34.862	0.783
	4	49.48	0.0001	19	0.73	44.703	0.770
Pb							
	1	6.49	0.02	13	0.35	4.271	0.041
	2	12.58	0.002	19	0.41	5.920	0.009
	3	17.97	0.0005	19	0.49	9.940	0.041
	4	12.86	0.002	19	0.42	11.495	0.038

predict that Cu will be bound preferentially over the other three trace elements,[95,96] and the strong affinity of POM for Cu has been shown elsewhere.[88,90] However, the PCF-o for Cu was not significantly higher than the PCF-r for Cu. It thus appears that in the Don River, the SPM-oxide pool is of greater importance in Cd, Zn, and Pb complexation than is the SPM-organic pool. The SPM-organic pool was only important for Cu complexation.

Organic substances can be of major importance in particulate trace metal interactions.[97-100] The pool of organic matter in natural environments is a complex and heterogeneous mixture drawn from a variety of possible sources. Yet, most attention has focused on high molecular weight organic substances, specifically humic substances.[55,101-103] While literature values of stability constants for metal-organic complexes are generally higher than are those for metal-inorganic complexes,[104] the majority of those stability complexes have been derived for humic substances. In systems where humic or humic-like substances are a minority of the organic pool, humic-derived constants may not be good predictors of the behavior of the organic pool. The greater importance of Don River SPM oxides in trace metal complexation, in comparison to the SPM-organic phase, has also been found for the Yukon and Amazon Rivers by Gibbs.[91]

E. Influence of Grain Size

If trace metal concentrations and/or geochemical associations are a predictable function of particle size, then the implication for remedial action is to focus on the smallest grain sizes in suspension. To investigate the nature of particle size-trace metal relationships in the Don, SPM samples from sites 2, 3, and 4 were physically size fractionated into five size classes (class 1, >102 μm; class 2, 63 to 102 μm, class 3, 10 to 63 μm, class 4, 5 to 10 μm; and class 5, 1 to 5 μm), using Nitex mesh sieves, and subsequently analyzed for geochemical fractionation of Cd, Cu, Zn, and Pb, using the same extraction procedure outlined above for nonsize fractionated samples.

If larger quantities of metal are concentrated in the smaller grain sized particles, based on surface to volume considerations, then the relative change in the surface area to volume ratio among size classes should be reflected in the relative concentrations of metal sorbed by those particle size classes. If particles are assumed to be spherical, then the surface area to volume ratio can be approximated using the formulas for the area and volume of a sphere, where the median particle diameter from each size class is used to approximate r. If a unit concentration of 1 is assumed for size class 1 (>102 μm), then the relative increase in metal concentrations expected, given the increase in surface area to volume ratio (SA:VOL) from size class 1 to size class 5 can be determined and compared to the actual increase in metal concentrations in the samples collected from the three sites. In addition, if a relationship between particle size and geochemical phase exists, then differences in geochemical partitioning among size classes should occur.

A comparison of actual relationships between particle size and trace metal concentrations with the theoretical relative increase predicted with decreasing particle size, shows little agreement (Figures 8 to 10). In addition, no consistent particle size-geochemical association relationships were evident, suggesting that discrete sediment geochemical phases are not reflected by specific particle sizes in the Don. Metal partitioning of size-fractionated samples shows similar patterns to nonsize-fractionated samples across size classes (Figures 4 to 10). Nonsize-fractionated results highlighted the importance of the oxides fraction for Cd, Zn, and Pb, and the POM fraction for Cu. These results suggest that oxide/organic coatings of particles across the entire size spectrum in the Don River occur, underlining the importance of these coatings in trace metal complexation reactions, as has been noted by other authors (e.g., Whitney[105]). Trace metal partitioning in the Don cannot be predicted as a function of particle size.

F. The Influence of Temperature and NaCl on Trace Metal Partitioning

In a recent study,[106] colder water temperatures and increased NaCl concentrations during winter months in the Don were found to be important factors influencing trace metal partitioning between particulate and solution phases. As temperature dropped from 10°C to 1°C, there was a significant decrease in the overall (bulk) partitioning of Cd, Zn, and Cu to the particulate pool (i.e., relatively more metal remained in the solution phase with lowered water temperatures). Increased NaCl concentrations, due to road-salt runoff, decreased

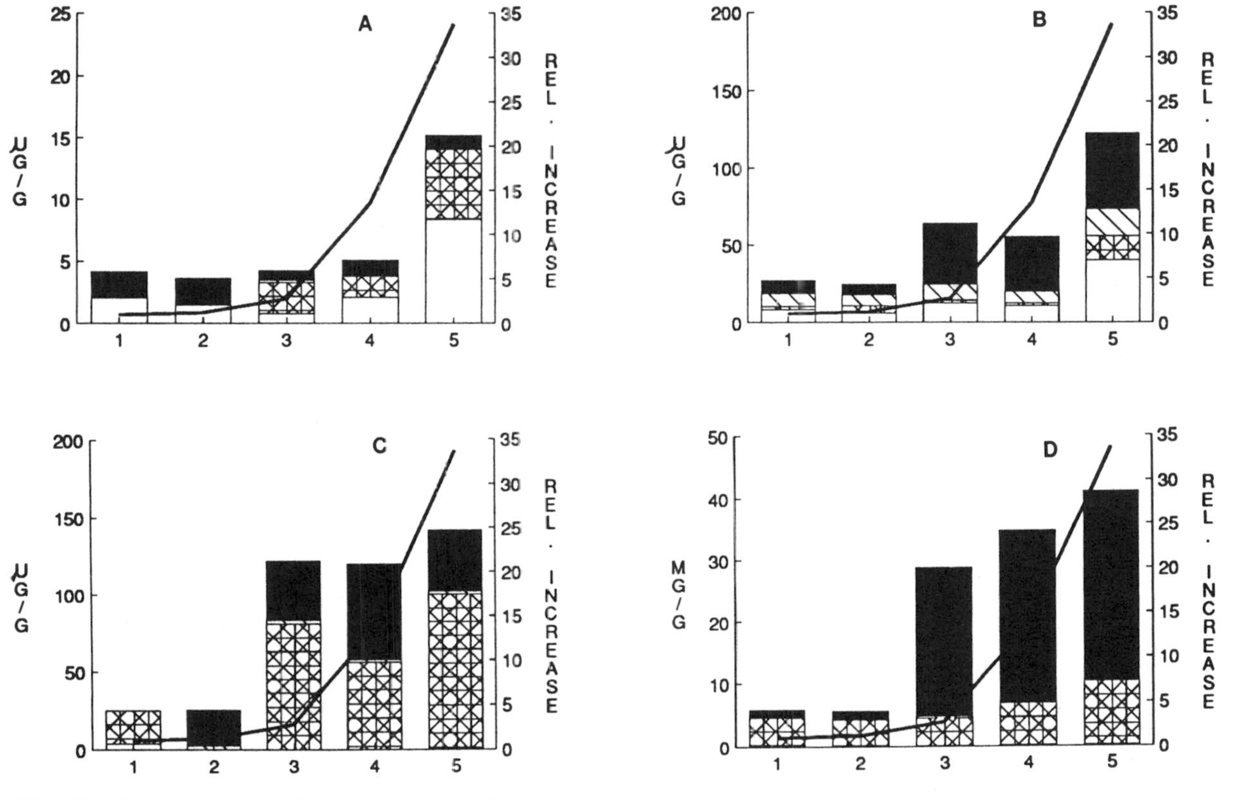

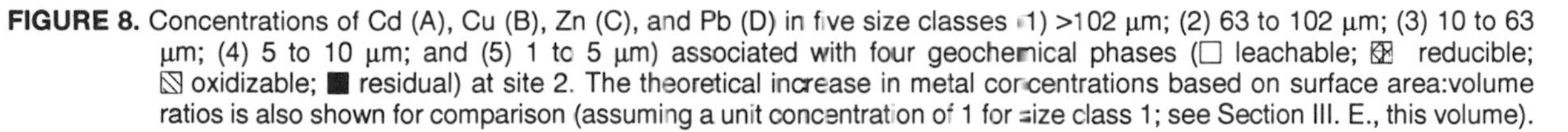

FIGURE 8. Concentrations of Cd (A), Cu (B), Zn (C), and Pb (D) in five size classes (1) >102 µm; (2) 63 to 102 µm; (3) 10 to 63 µm; (4) 5 to 10 µm; and (5) 1 to 5 µm) associated with four geochemical phases (□ leachable; ⊠ reducible; ▧ oxidizable; ■ residual) at site 2. The theoretical increase in metal concentrations based on surface area:volume ratios is also shown for comparison (assuming a unit concentration of 1 for size class 1; see Section III. E., this volume).

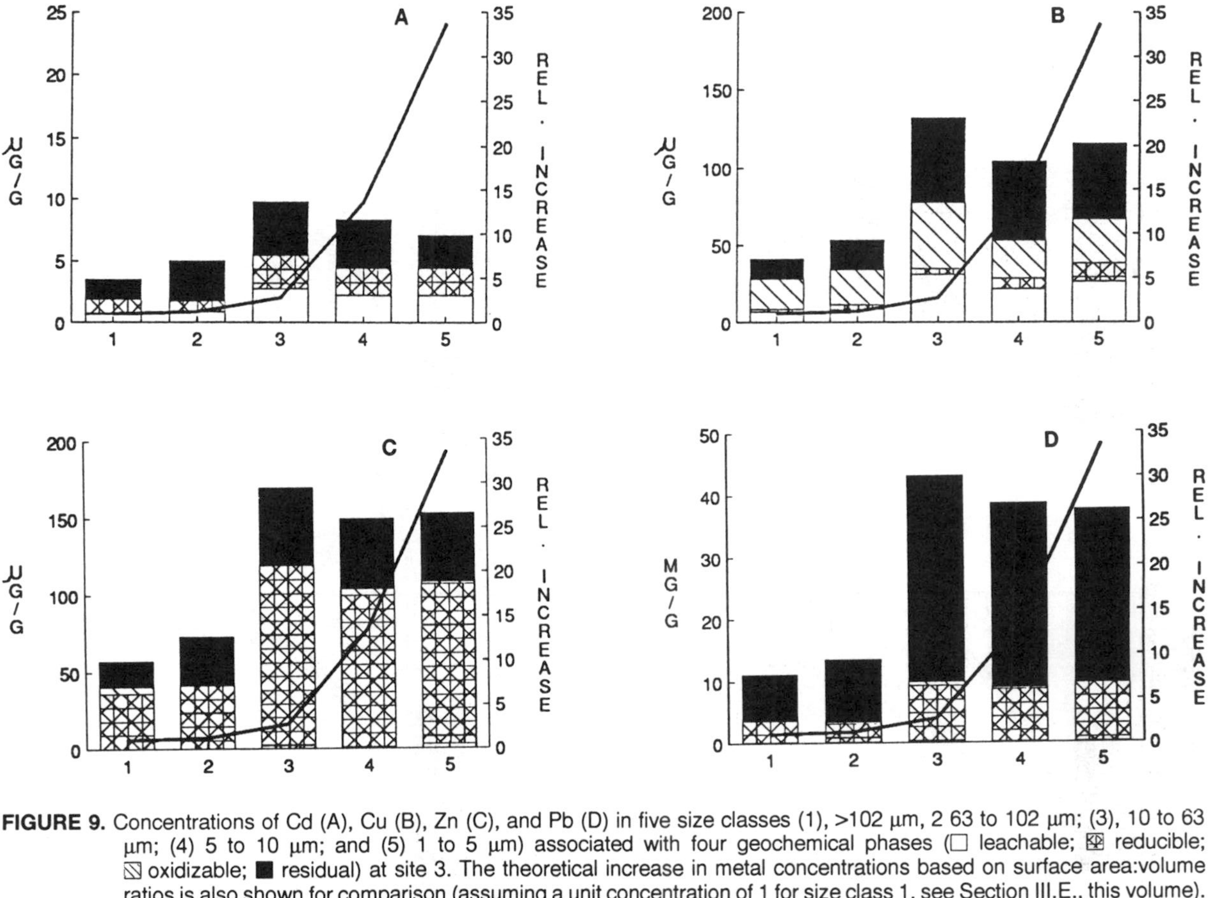

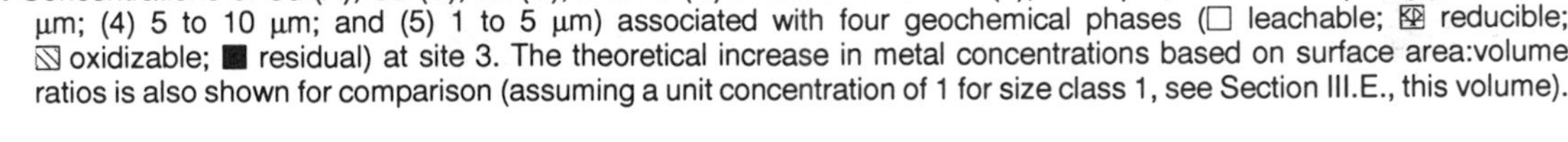

FIGURE 9. Concentrations of Cd (A), Cu (B), Zn (C), and Pb (D) in five size classes (1), >102 µm, 2 63 to 102 µm; (3), 10 to 63 µm; (4) 5 to 10 µm; and (5) 1 to 5 µm) associated with four geochemical phases (□ leachable; ⊠ reducible; ▧ oxidizable; ■ residual) at site 3. The theoretical increase in metal concentrations based on surface area:volume ratios is also shown for comparison (assuming a unit concentration of 1 for size class 1, see Section III.E., this volume).

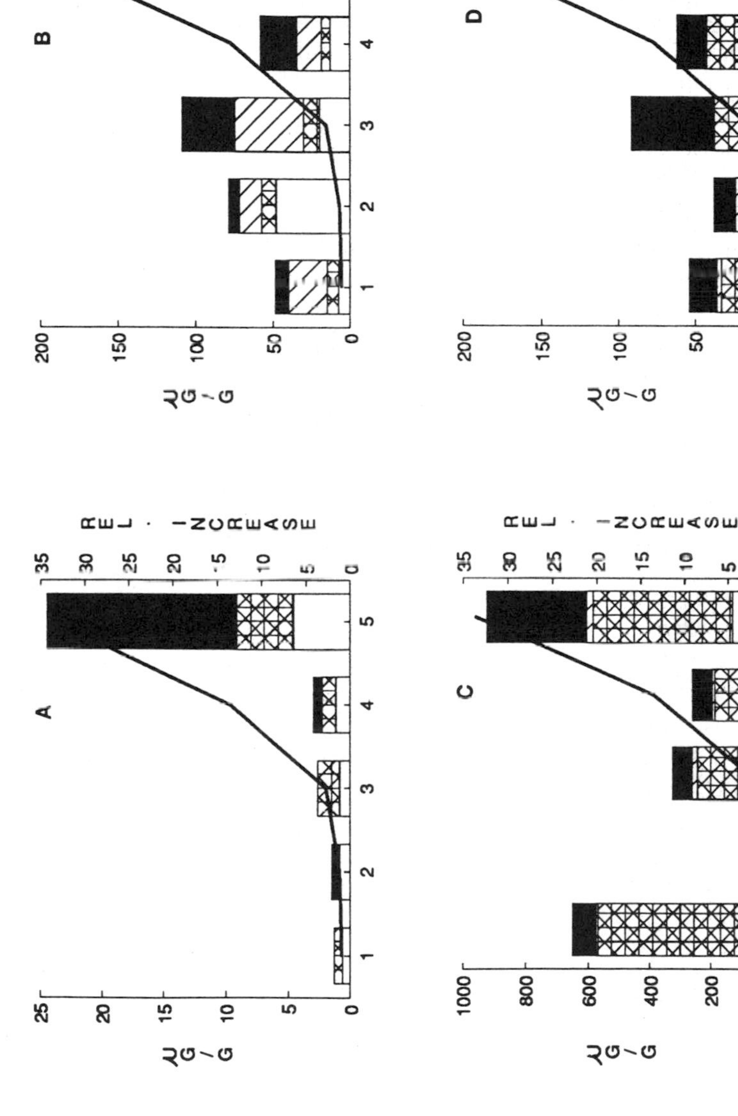

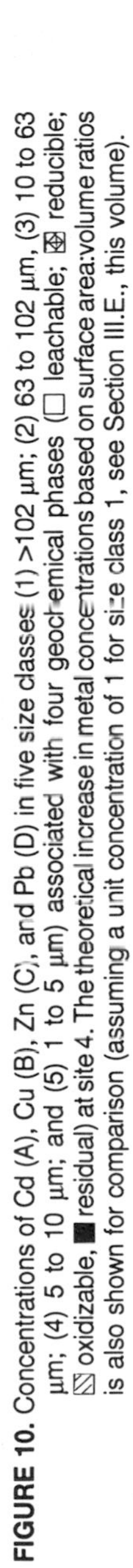

FIGURE 10. Concentrations of Cd (A), Cu (B), Zn (C), and Pb (D) in five size classes (1) >102 µm; (2) 63 to 102 µm, (3) 10 to 63 µm; (4) 5 to 10 µm; and (5) 1 to 5 µm) associated with four geochemical phases (□ leachable; ⊠ reducible; ▨ oxidizable, ■ residual) at site 4. The theoretical increase in metal concentrations based on surface area:volume ratios is also shown for comparison (assuming a unit concentration of 1 for size class 1, see Section III.E., this volume).

the partitioning of Cd and Zn to the leachable and oxidizable particulate pools specifically. Again, the relationships were negative: relatively less Cd and Zn partitioned to the particulate pools with increasing NaCl concentrations. The partitioning of Cu to the leachable particulate pool was decreased by decreasing water temperature. Partitioning of Cu to the oxidizable particulate pool was decreased by increasing concentrations of both Ca^{2+} and NaCl. Thus, during winter months in the Don River, relatively more trace metal remains in the dissolved and potentially more bioavailable pool due to temperature and NaCl effects. The partitioning of all three cations to the reducible fraction of the SPM was found to be independent of the environmental variables examined in the study.

IV. CONCLUSIONS

Sorption processes (adsorption, absorption, precipitation) are the principal mechanisms governing reactions between trace metals and particulates in aquatic systems. SPM plays a key role in determining the complexation, transport, and ultimate fate of trace metals. In the SPM pool, Fe and Mn oxides and POM are important geochemical fractions involved in trace metal particulate associations. Carrier particles (e.g., clays), are often coated in natural systems with a surface film of oxides and/or POM. Thus, particulate trace metal complexation is essentially controlled by sorption reactions between the adsorbed layer of oxides and/or POM and trace metals at the solid solution interface. Sorption reactions are influenced by a variety of factors, such as pH, temperature, pE, sedimentary geochemistry, types and concentrations of ligands, and trace metals.

Most of our current knowledge on aqueous surface chemistry has been derived from laboratory studies where system parameters can be accurately described and controlled. More field studies that apply current theory e.g., Tessier et al.[12] are essential prerequisites for determining the applicability of such information for natural systems. Such studies, that ask the right questions and measure the appropriate parameters, will greatly advance the field of aqueous trace metal complexation.

A case study of the Don River provides useful insights into trace metal particulate interactions in a real system. Absolute concentrations of Cd, Cu, Zn, and Pb were found to be both spatially and temporally variable in the Don River. However, geochemical partitioning of each cation was found to be relatively stable. Patterns of geochemical association suggest an underlying order to trace metal particulate associations that may be ultimately modeled. The reducible SPM fraction was found to be the major particulate transport vector for Cd, Zn, and Pb. Cu was found predominantly associated with the oxidizable SPM fraction. A comparison of PCF values for the two sediment fractions revealed a significantly greater scavenging ability of the reducible SPM pool for Cd, Zn, and Pb. No significant differences between the two pools emerged for Cu. In the Don, the SPM-reducible pool is of greater importance in particulate trace metal complexation than is the oxidizable fraction.

Results from size-fractionated samples did not show a consistent increase in trace metal concentrations with decreasing particle size, as theoretical surface area to volume considerations would predict. Nor did any consistent particle size-geochemical partitioning relationships emerge, suggesting that specific particle sizes do not represent any discrete geochemical phase. Geochemical partitioning was similar to nonsize-fractionated samples, emphasizing the oxide and POM fractions, suggesting that oxide/organic coatings of particles probably occurs across the particle spectrum in this system. Trace metal concentrations and geochemical partitioning cannot be predicted as a function of particle size without accounting for the degree and nature of particle coatings.

Lowered water temperature and increased NaCl concentrations during winter months in the Don River were found to decrease the relative partitioning of Cd, Zn, and Cu to the particulate pool, increasing the solution concentrations of these cations. Decreasing temperature decreased the partitioning to the total particulate pool from the solution phase, while NaCl decreased the partitioning to the oxidizable particulate phase. Thus, during winter months in the Don, relatively more metal remains in the solution and potentially more bioavailable pool. Partitioning between the particulate-reducible pool (the major particulate pool for trace metal complexation in this system) and the solution phase was independent of any of the environmental parameters investigated.

REFERENCES

1. Honeyman, B. D. and Santschi, P. H., Metals in aquatic systems: predicting their scavenging residence times from laboratory data remains a challenge, *Environ. Sci. Technol.*, 22, 862, 1988.
2. Scrudato, R. J. and Hocutt, G., An *in-situ* integrated suspended sediment stream sampler (IS3), *Environ. Geol. Water Sci.*, 12; 177, 1988.
3. Sheintuch, M. and Rebhun, M., Adsorption isotherms for multisolute systems with known and unknown composition, Water Res., 22, 421, 1988.
4. Di Toro, D. M., Mahony, J. D., Kirchgraber, P. R., O'Bryne, A. L., Pasquale, L. R., and Piccirilli, D. C., 1986. Effects of nonreversibility, particle concentration, and ionic strength on heavy metal sorption, *Environ. Sci. Technol.*, 20(1),55, 1986.
5. Hart, B. T., Uptake of trace metals by sediments and suspended particulates: a review, *Hydrobiology,* 91, 299, 1982.
6. Jaquet, J. M., Davaud, E., Rapin, F., and Vernet, J. P., Basic concepts and associated statistical methodology in the geochemical study of lake sediments, *Hydrobiology,* 91, 139, 1982.
7. Ongley, E. D., Bynoe, M. C., and Percival, J. B., Physical and geochemical characteristics of suspended solids, Wilton Creek, Ontario, *Hydrobiology,* 91; 41, 1982.

8. Benjamin, M. M. and Leckie, J. O., Adsorption of metals at oxide interfaces: effects of the concentrations of adsorbate and competing metals, in *Contaminants and Sediments, Vol. 2,* Baker, R. A., Ed., Ann Arbor Science, Ann Arbor, MI, 1980, 305.
9. O'Connor, D. J. and Connolly, J. P., The effect of concentration of adsorbing solids on the partition coefficient, *Water Res.,* 14, 1517, 1980.
10. Sinclair, P., Beckett, R., and Hart, B. T., Trace elements in suspended particulate matter from the Yarra River, Australia, *Hydrobiology,* 176/177, 239, 1989.
11. Bowers, A. R. and Huang, C. P., Role of Fe(III) in metal complex adsorption by hydrous solids, *Water Res.,* 21(7), 757, 1987.
12. Tessier, A., Rapin, F., and Carignan, R., Trace metals in oxic lake sediments: possible adsorption onto iron oxyhydroxides, *Geochim. Cosmochim. Acta,* 49, 183, 1985.
13. Luoma, S. N., Can we determine the biological availability of sediment-bound trace elements? *Hydrobiology,* 176/177, 379, 1989.
14. Uchrin, C. G. and Weber, W. J., Modeling of transport processes for suspended solids and associated pollutants in river-harbor-lake systems, in *Contaminants and Sediments,* Vol. 1, Baker, R. A. Ed., Ann Arbor Science, Ann Arbor, MI, 1980, 407.
15. Davis, J. A. and Leckie, J. O., Effect of adsorbed complexing ligands on trace metal uptake by hydrous oxides, *Environ. Sci. Technol.,* 12(12), 1309, 1978.
16. Horowitz, A. J., Rinella, F., Lamothe, P., Miller, T., Edwards, T., Roche, R., and Rickert, D., Variations in suspended sediment and associated trace element concentrations in selected riverine cross sections, *Environ. Sci. Technol.,* 24, 1313, 1990.
17. McIlroy, L. M., DePinto, J. V., Young, T. C., and Martin, S. C., Partitioning of heavy metals to suspended solids of the Flint River, Michigan, *Environ. Toxicol. and Chem.,* 5, 609, 1986.
18. Rendell, P. S. and Batley, G. E., Adsorption as a control of metal concentrations in sediment extracts, *Environ. Sci. Technol.,* 14(3), 314, 1980.
19. Allard, B., Hakansson, K., and Karlsson, S., The importance of sorption phenomena in relation to trace element speciation and mobility, in *Speciation of Metals in Water, Sediment and Soil Systems,* Lecture Notes in Earth Sciences #11, Springer-Verlag, Berlin, 1987.
20. Forstner, U., Metal speciation in solid wastes — factors affecting mobility, in *Speciation of Metals in Water, Sediment and Soil Systems,* Lecture Notes in Earth Sciences #11, Springer-Verlag, Berlin, 1987.
21. Schindler, P. W., Surface complexes at oxide-water interfaces, in *Adsorption of Inorganics at Solid-Liquid Interfaces,* Anderson M. A. and Rubin, A. J., Eds., Ann Arbor Science, Ann Arbor, MI, 1981.
22. Leckie, J. O. and James, J., Control mechanisms for trace metals in natural waters, in *Adsorption of Inorganics at Solid-Liquid Interfaces,* Anderson, M. A. and Rubin, A. J., Ann Arbor Science, Ann Arbor, MI, 1974.
23. Sigg, L., Surface chemical aspects of the distribution and fate of metal ions in lakes, in *Aquatic Surface Chemistry,* Stumm, W., E., John Wiley & Sons, New York, 1987.
24. Westall, J. C., Adsorption mechanisms in aquatic surface chemistry, in *Aquatic Surface Chemistry,* Stumm, W., Ed., John Wiley & Sons, New York, 1987.

25. Benjamin, M. M. and Leckie, J. O., Multiple-site adsorption of Cd, Cu, Zn, and Pb on amorphous iron oxyhydroxide, *J. Colloid Interface Sci.,* 79 (1), 209, 1981.
26. Vuceta, J. and Morgan, J.J., Chemical modeling of trace metals in fresh waters: role of complexation and adsorption, *Environ. Sci. Technol.,* 12(12), 1303, 1978.
27. Tessier, A., Carignan, R., Dubreuil, B., and Rapin, F., Partitioning of zinc between the water column and the oxic sediments in lakes, *Geochim. Cosmochim. Acta,* 53, 1511, 1989.
28. Forstner, U. and Wittmann, G. T. W., *Metal Pollution in the Aquatic Environment,* 2nd ed., Springer-Verlag, New York, 1981.
29. Andelman, J. B., Incidence, variability and controlling factors for trace elements in natural, fresh waters, in *Trace Metals and Metal-Organic Interactions in Natural Waters,* Singer, P. C., Ed., Ann Arbor Science, Ann Arbor, MI, 1974.
30. Parks, G. A., Aqueous surface chemistry of oxides and complex oxide minerals. Isoelectric point and zero point of charge, in *Equilibrium Concepts in Natural Water Systems. Symp. ACS #67,* American Chemical Society, Washington, D.C., 1967.
31. Bruemmer, G. W., Gerth, J., and Tiller, K. G., Reaction kinetics of the adsorption and desorption of nickel, zinc and cadmium by goethite. I. Adsorption and diffusion of metals, *J. Soil Sci.,* 39, 37, 1988.
32. Papadopoulos, P. and Rowell, D. L., The reactions of cadmium with calcium carbonate surfaces, *J. Soil Sci.,* 39, 23, 1988.
33. Schindler, P. W., Surface complexes at oxide-water interfaces, in *Adsorption of Inorganics at Solid-Liquid Interfaces,* Anderson, M. A. and Rubin, A. J., Eds., Ann Arbor Science, Michigan, 1981, 1.
34. Dzombak, D. A. and Morel, F. M. M., Sorption on hydrous ferric oxide at high sorbate/sorbent ratios: equilibrium, kinetics and modeling, *J. Colloid Interface Sci.,* 112(2), 588, 1986.
35. Westall, J. C., Reactions at the oxide-solution interface: chemical and electrostatic models, in *Geochemical Processes at Mineral Surfaces. Symp. ACS #323,* Davis, J. A. and Hayes, K. F., Eds., American Chemical Society, Washington, D.C., 1986.
36. Farley, K. J., Dzombak, D. A., and Morel, F. M. M., A surface precipitation model for the sorption of cations on metal oxides, *J. Colloid Interface Sci.,* 106(1), 226, 1985.
37. Balistrieri, L. S. and Murray, J. W., Metal-solid interactions in the marine environment: estimating apparent equilibrium binding constants, *Geochim. Cosmochim. Acta,* 47, 1091, 1983.
38. Jenne, E. A., Controls on Mn, Fe, Co, Ni, Cu, and Zn concentrations in soils and water: the significant role of hydrous Mn and Fe oxides, in *Trace Inorganics in Water,* Baker, R. A., Ed., American Chemical Society Publication #73, Washington, D.C., 1968.
39. Benjamin, M. M. and Leckie, J. O., Conceptual model for metal ligand-surface interactions during adsorption, *Environ. Sci. Technol.,* 15(9), 1050, 1981.
40. Davis, J. A., Adsorption of natural dissolved organic matter at the oxide/water interface, *Geochim. Cosmochim. Acta,* 46, 2381, 1982.
41. Lion, L. W., Altmann, R. S., and Leckie, J. O., Trace metal adsorption characteristics of estuarine particulate matter: evaluation of contributions of Fe/Mn oxide and organic surface coatings, *Environ. Sci. Technol.,* 16, 660, 1982.

42. Tipping, E., The adsorption of aquatic humic substances by iron oxides, *Geochim. Cosmochim. Acta,* 45, 191, 1981.
43. Tipping, E., Some aspects of the interactions between particulate oxides and aquatic humic substances, *Mar. Chem.,* 18, 161, 1986.
44. Davis J. A., Complexation of trace metals by adsorbed natural organic matter, *Geochim. Cosmochim. Acta,* 48, 679, 1984.
45. Turner, D. R., Varney, M. S., Whitfield, M., Mantoura, R. F. C., and Riley J. P., Electrochemical studies of copper and lead complexation by fulvic acid. I. Potentiometric measurements and a critical comparison of metal binding models, *Geochim. Cosmochim. Acta,* 50, 289, 1986.
46. Laxen, D. P. H., Trace metal adsorption/coprecipitation on hydrous ferric oxide under realistic conditions, *Water Res.,* 19(10), 1229, 1985.
47. Buffle, J., Tessier, A., and Haerdi, W., Interpretation of trace metal complexation by aquatic organic matter, in *Complexation of Trace Metals in Natural Waters,* Kramer, C. J. M. and Duinker, J. C., Eds., The Junk Publ., The Hague, 1984.
48. Frimmel, F. H., Immerz, A., and Niedermann, H., Complexation capacities of humic substances isolated from freshwater with respect to copper (II), mercury (II) and iron (II, III), in *Complexation of Trace Metals in Natural Waters,* Kramer, C. J. M. and Duinker, J. C., Eds., Junk Publ., The Hague, 1984.
49. Raspor, B., Nurnberg, H. W., Valenta, P., and Branica, M., Significance of dissolved humic substances for heavy metal speciation in natural water, in *Complexation of Trace Metals in Natural Waters,* Kramer, C. J. M. and Duinker, Eds., Junk Publ., The Hague, 1984.
50. Hansen, A. M., Leckie, J. O., Mandelli, E. F., and Altmann, R. S., Study of copper (II) association with dissolved organic matter in surface waters of three Mexican coastal lagoons, *Environ. Sci. Technol.,* 24(5), 683, 1990.
51. Han-Bin, X., Stumm, W., and Sigg, L., The binding of heavy metals to algal surfaces, *Water Res.,* 22(7), 917, 1988.
52. Altmann, R. S. and Buffle, J., The use of differential equilibrium functions for interpretation of meta binding in complex ligand systems: its relation to site occupation and site affinity distributions, *Geochim. Cosmochim. Acta,* 52, 1505, 1988.
53. Van De Meent, D., De Leeuw, J. W., Schenck, P. A., and Salomons, W., Geochemistry of supended particulate matter in two natural sedimentation basins of the River Rhine, *Water Res.,* 19(11), 1333, 1985.
54. Benjamin, M. M. and Leckie, J. O., Competitive adsorption of Cd, Cu, Zn, and Pb on amorphous iron oxyhydroxide, *J. Colloid Interface Sci.,* 83(2), 410, 1981.
55. Lovgren, L. and Sjoberg, J., Equilibrium approaches to natural water systems — 7. Complexation reactions of copper (II), cadmium (II) and mercury (II) with dissolved organic matter in a concentrated bog-water, *Water Res.,* 23(3), 327, 1989.
56. Martell, A. E., Motekaitis, R. J., and Smith, R. M., Structure-stability relationships of metal complexes and metal speciation in environmental aqueous solutions, *Environ. Toxic Chem.,* 7, 414, 1988.
57. Dzombak, D. A., Fish, W., and Morel, F. M. M., Metal-humate interactions. 1. Discrete ligand and continuous distribution models, *Environ. Sci. Technol.,* 20(7), 669, 1986.
58. Luoma, S. N., Bioavailability of trace metals to aquatic organisms — a review, *Sci. Total Environ.,* 28 1, 1983.

59. Feijtel, T. C., DeLaune, R. D., and Patrick, W. H., Biogeochemical control on metal distribution and accumulation in Louisiana sediments, *J. Environ. Qual.*, 17(1), 88, 1988.
60. Turner, D. R., Whitfield, M., and Dickson, A. G., The equilibrium speciation of dissolved components in freshwater and seawater at 25°C and 1 atm pressure, *Geochim. Cosmochim. Acta*, 45, 855, 1981.
61. Walling, D. E. and Moorehead, P. W., The particle size characteristics of fluvial suspended sediment: and overview, *Hydrobiology*, 176/177, 125, 1989.
62. Beckett, R., Nicholson, G., Hart, B. T., Hansen, M., and Giddings, J. C., Separation and size characterization of colloidal particles in river water by sedimentation field-flow fractionation, *Water Res.*, 22(12), 1535, 1988.
63. Jenne, E. A., Kennedy, V. C., Burchard, J. M., and Ball, J. W., Sediment collection and processing for selective extraction and for total trace element analyses, in *Contaminants and Sediments*, Vol. 2, Baker, R. A., Ed., Ann Arbor Science, Ann Arbor, MI, 1980, 169.
64. Ziegler, C. K. and Lick W., The transport of fine-grained sediments in shallow waters, *Environ. Geol. Water Sci.*, 11(1), 123, 1988.
65. Mudroch, A. and Duncan, G. A., Distribution of metals in different size fractions of sediment from the Niagara River, *J. Great Lakes Res.*, 12(2), 117, 1986.
66. Jennett, J. C., Effler, S. W., and Wixson, B. G., Mobilization and toxicological aspects of sedimentary contaminants, in *Contaminants and Sediments*, Vol. 1, Baker, R. A., Ed., Ann Arbor Science, Ann Arbor, MI, 1980, 429.
67. Dzombak, D. A. and Morel, F. M. M., *Surface Complexation Modeling. Hydrous Ferric Oxide*, John Wiley & Sons, New York, 1990.
68. Horowitz, A. J., *Sediment-Trace Element Chemistry*, 2nd ed., Lewis Publishers, Chelsea, MI, 1991.
69. Warren, L. A. and Zimmerman, A. P., Rain event associated changes in metal transport by suspended sediments in the Don River, *Verh. Internat. Verein. Limnol.*, 24, 2235, 1991.
70. Zinder, B., Furrer, G., and Stumm, W., The coordination chemistry of weathering II. Dissolution of Fe(III) oxides. *Geochim. Cosmochim. Acta*, 50, 1986.
71. Furrer, G. and Stumm, W., The coordination chemistry of weathering. I. Dissolution kinetics of delta-Al2O3 and BeO, *Geochim. Cosmochim. Acta*, 50, 1847, 1986.
72. Bleam, W. F. and McBride, M. B., Cluster formation versus isolated site adsorption. A study of Mn (II) and Mg (II) adsorption on Boehmite and Goethite, *J. Colloid and Interface Sci.*, 103, 124, 1985.
73. Campbell, P. G. C., Lewis, A. G., Chapman, A. A., Crowder, A. A., Fletche, W. K., Imber, B., Luoma, S. N., Stokes, P. M., and Winfrey, M., *Biologically Available Metals in Sediments*. NRCC Report No. 276943, National Research Council of Canada, Ottawa, 1988.
74. Calmano, W. and Forstner, U., Chemical extraction of heavy metals in polluted river sediments in central Europe, *Science Tot. Environ.*, 28, 7790, 1983.
75. Meguellati, N., Robbe, D., Marchandise, P., and Astruc, M., A new chemical extraction procedure in the fractionation of heavy metals in sediments — interpretation, in *Proc. Int. Conf, Heavy Metals Environment*, CEP Consultants, Edinburgh, U. K., 1983, 1090.

76. Badri, M. A. and Aston, S. R., A comparative study of sequential extraction procedures in the geochemical fractionation of heavy metals in estuarine sediments, in *Proc. Int. Conf. Heavy Metals Environment,* CEP Consultants, Edinburgh, U.K., 1981, 705.
77. Hoffman, S. J. and Fletcher, W. K., Organic matter scavenging of copper, zinc, molybdenum, iron and manganese, estimated by sodium hypochlorite extraction (pH 9.5), *J. Geochem. Explor.,* 15, 549, 1980.
78. Tessier, A., Campbell, P. G. C., and Bisson, M., Sequential extraction procedure for the speciation of particulate trace metals, *Anal. Chem.,* 51, 844, 1979.
79. Kheboian, C. and Bauer, C. F., Accuracy of selective extraction procedures for metal speciation in model aquatic sediments, *Anal. Chem.,* 59, 1417, 1987.
80. Nirel, P. M. V. and Morel, F. M. M., Pitfalls of sequential extractions, *Water Res.,* 24, 1055, 1990.
81. Belzille, N., Lecomte, P., and Tessier, A., Testing readsorption of trace elements during partial chemical extractions of bottom sediments, *Environ. Sci Technol.,* 23, 1015, 1989.
82. Campbell, P. G. C. and Tessier, A., Current status of metal speciation studies, in *Metals Speciation, Separation and Recovery,* Patterson, J. W. and Passino, R., Eds., Lewis Publishers, Chelsea, MI, 1987.
83. Guy, R. D., Chakrabarti, C. L., and Schramm, L. L., The application of a simple chemical model of natural waters to metal fixation in particulate matter, *Can. J. Chem.,* 53, 661, 1975.
84. Bendell-Young, L. I. and Harvey, H. H., The relative importance of manganese and iron oxides and organic matter in the sorption of trace metals by surficial lake sediments *Geochim. Cosmochim. Acta,* 56, 1175, 1992.
85. Boyd, D., Fluvial sedimentation in a small urban river: The Don, University of Tornoto, Ontario, Canada, Dept. of Geography, unpublished data.
86. Chapman, L. J. and Putnam, D. F., *The Physiography of Southern Ontario,* University of Toronto Press, Toronto, Ontario, Canada, 1966.
87. Hoffman, D. W. and Richards, N. R., Soil Survey of York County, Report No. 19 Ontario Soil Survey, Experimental Farms Service, Guelph Ontario, Canada, 1955.
88. Pardo, R., Barrado, E., Perez, L., and Vega, M., Determination and speciation of heavy metals in sediments of the Pisuerga River, *Water Res.,* 24, 373, 1990.
89. Chen, J., Dong, L., and Deng, B., A study on heavy metal partitioning in sediments from Poyang Lake in China, *Hydrobiology,* 176, 159, 1989.
90. Tessier, A., Campbell, P. G. C., and Bisson, M., Trace metal speciation in the Yamaska and St. Francois Rivers (Quebec), *Can. J. Earth Sci.,* 17, 90, 1980.
91. Gibbs, R. J., Transport phases of transition metals in the Amazon and Yukon Rivers, *Geol. Soc. Am. Bull.,* 88, 829, 1977.
92. Bendell-Young, L. I., Accumulation of Metals by Invertebrates and Fish, Ph.D. dissertation, University of Toronto, Ontario, Canada, 1990.
93. Luoma, S. N. and Bryan, G. W., A statistical assessment of the form of trace metals in oxidized estuarine sediments employing chemical extractants, *Sci. Total Environ.,* 17, 165, 1981.
94. Forstner, U., Chemical forms and reactivities of metals in sediments, *Chemical Methods for Assessing Bioavailable Metals in Sludges and Soils,* Leschber, R., Davis, R. D., and L'Hermite, P. O., Eds., Elsevier, London, 1985, 1.
95. Smith, R. M. and Martell, A. E., *Critical Stability Constants,* Plenum Press, New York, 1976.

96. Stevenson, F. J., Stability constants of Cu^{2+}, Pb^{2+}, and Cd^{2+} complexes with humic acids, *Soil Sci. Soc. Am. J.,* 40, 665, 1976.
97. Wangersky, P. J., Biological control of trace metal residence time and speciation: a review and synthesis, *Mar. Chem.,* 18, 269, 1986.
98. Hargitai, L., The role of humus status of soils in binding toxic elements and compounds, *Sci. Total Environ.,* 81/82, 643, 1989.
99. Weber, J. H., Binding and transport of metals by humic materials, in *Humic Substances and Their Role in the Environment,* Frimmel, F. H., and Christman, R. F., Eds., John Wiley & Sons New York, 1988, 165.
100. Schnitzer, M., Soil organic matter — the next 75 years, *Soil Sci.,* 151, 41, 1991.
101. Hering, J. G. and Morel, F. M. M., Kinetics of trace metal complexation: role of alkaline-earth metals, *Environ. Sci. Technol.,* 22, 469, 1988.
102. Fish, W., Dzombak, D. A., and Morel, F. M. M., Metal-humate interactions. 2. Application and comparison of models, *Environ. Sci. Technol.,* 20, 676, 1986.
103. Perdue, E. M. and Lytle, C. R., Distribution model for binding of protons and metal ions by humic substances, *Environ. Sci. Technol.,* 17, 654, 1983.
104. Reuter, J. H. and Perdue, E. M., Importance of heavy metal-organic matter interactions in natural waters, *Geochim. Cosmochim. Acta,* 41, 325, 1977.
105. Whitney, P. R., Relationship of manganese-iron oxides and associated heavy metals to grain size in stream sediment, *J. Geochem. Explor.,* 4, 251, 1975.
106. Warren, L. A. and Zimmerman, A. P., The influence of temperature and NaCl on Cd, Cu and Zn partitioning among suspended particulate and dissolved phases in an urban river, *Geochim. Cosmochim. Acta,* in review.
107. Sigg, L., Stumm, W., and Zinder, B., Chemical processes at the particle-water interface; implications concerning the form of occurrence of solute and adsorbed species, in *Complexation of Trace Metals in Natural Waters,* Kramer, C. J. M. and Duinker, J. C., Eds., Junk Publ., The Hague, 1984.

CHAPTER 6

Ecotoxicological Implications of Fluvial Suspended Particulates

Salem S. Rao, Berand J. Dutka, and Colin M. Taylor

TABLE OF CONTENTS

0-87371-678-7/93/$0.00+$.50

I. INTRODUCTION

In order to better understand and predict the fate of contaminants moving through the aquatic ecosystem, more basic research is required. The role that suspended inorganic particles, such as clays and silts, play in the adsorption and subsequent transportation of contaminants and toxicants within the aquatic ecosystem is well recognized.[1] However, there are still many knowledge gaps that need to be addressed with regard to the role that suspended particulates (biological-inorganic material complexes) play in the concentration, biotransformation, and transportation of contaminants. To more thoroughly understand this phenomenon, we need to better understand, among many factors, the formation of suspended particulates, the role bacteria play in the formation of these complexes, and the partitioning behavior of the various contaminants/toxicants between solid and liquid phase.

The sorption of contaminants/toxicants onto suspended particulates has been shown to be influenced by a number of factors such as: particle size, concentration and spacial distribution of inorganic and organic particles, and bacterial densities.[1] Bacterial reproduction rates and the rate of bacterial adhesion to surfaces are key elements in the formation and behavior of contaminant-bacterial complexes.[2-6]

Bacteria have the highest surface area to volume ratio of all cells. Consequently, they have the potential for extensive interactions with their aqueous surroundings, which allows them to have a high capacity to sorb large quantities of toxicants onto their surfaces.[5] This effectively immobilizes and removes many aquatic contaminants out of the aquatic system.[5,6] Another important process in the transport of chemical contaminants in natural waters is the partitioning which occurs between the operationally defined "dissolved phase" and reservoirs of organics found in "compartments" such as suspended particulates.[7] Furthermore, it has been indicated[7] that the main factor governing the adsorption of lipophilic contaminants to suspended solids is the organic carbon content of these suspended solids. The fundamental surface properties of bacteria, which promote aggregates, deserve greater consideration, since naturally occurring aggregates act as a source of nutrients for bacteria, which in turn affects water quality.[5] Particle size and content may influence suspended particulate-toxicant interactions, and, thus, these complexes may be important vehicles for long-range transport of soluble and adsorbed contaminants.

To gain a better understanding of the partitioning behavior of some contaminants between the liquid phase and suspended particulates, contaminant adsorption studies were initiated using water-soluble dyes and suspended particulates collected from a highly polluted part of the Yamaska River, in the Canadian province of Quebec. Two dyes (Acid Orange-60 and Basic Violet-1) were chosen as model contaminants for this study because of their known prevalence in these waters, as a result of industrial discharges.

In this chapter, we present data from an investigation to ascertain the role that suspended bacterial aggregates play in the contaminant-adsorption process. Also in this chapter, we compare the composition of various suspended particulates from other Canadian rivers and discuss the implications of these observations.

In the following discussion, for clarity, suspended particulates are defined as bacteria-inorganic particle complexes. Some of the contents of this chapter are extracted from previously published material.[8]

II. MATERIALS AND METHODS

A. Sampling Site

The Yamaska River, which is situated in the Eastern Townships of Quebec about 70 km east of Montreal, is a tributary of the St. Lawrence River. This river has been highly polluted by discharges from textile industries, domestic effluents, farm waste, and land run-off (pesticides and herbicides), and, as a result, the water quality in this basin has been generally regarded as poor.[9,10]

During the summer of 1988, water samples (200 l) were collected from this river and maintained at 20°C until all biochemical analyses and contaminant adsorption studies were completed. Later, samples from other rivers (Thames River, Ontario; Athabasca River, Alberta; St. Maurice River, Quebec) were also collected and characterized to determine relationships, if any, between particle-size distribution patterns and bacterial concentrations.

B. Particle-Size Distribution Analysis

The sizes of the suspended particulates and the distribution pattern of these sizes (ranging between 1 and 100 μm), was determined by a Malvern series 2600 Laser Diffraction Particle Sizer.[11] This system uses a light source containing a 3 mV laser, a receiving optics assembly, and an electronic circuitry which interfaces with a microcomputer. The distribution of the various sized particles in the sample was derived from measurements of the near-forward Fraunhofer diffraction spectrum that is provided by particle groups randomly distributed in a sample cell mounted in a beam path between the laser source and the detector array. Since suspended particulate matter was found in the transported Yamaska River samples, it was assumed that these flocculated aggregates were physically stable between the time of collection and the time of analysis. This assumption was based on microscopic observations of floc stability and longevity under conditions of moderate agitation.[12]

C. Fractionation of Suspended Particulates

The various sizes of suspended particulates were separated from a 5-l water sample following the modified cascade filtration procedure of Rao and Kwan[13] which uses the following sized filters: 88, 64, 40, 20, 8, and 3 μm. The 8-μm and the 3-μm filters were both made of polycarbonate, while the other filters were made of Nitex®-3 (100% nylon polyamide). Extreme care was taken in handling the various filters to minimize particle disintegration. Filter clogging was reduced by gently mixing the sample during filtration and subsequent resuspension of the material. The suspended particulates from each of the sieves were carefully resuspended into 200 ml of distilled water and analyzed for bacteria, microtox toxicity, organic carbon, and nitrogen content.

D. Bacterial Analysis

A 1-ml sample from each of the various sized groups of suspended particulates was diluted to 10 ml with sterile, low-response water and was then homogenized using a vortex mixer set at its highest speed for 1 min in order to separate bacteria from the inorganic matrix.[14] This mixture was immediately subjected to membrane filtration using a 0.2-μm Nuclepore® membrane filter and then staining the bacteria with Acridine Orange for 3 min. Total bacterial (Acridine Orange Direct Counts) concentration was determined using fluorescent microscopy.[15]

E. Organic Carbon and Nitrogen

Particulate organic carbon and nitrogen analysis were performed according to Standard Methods.[16]

F. Microtox Toxicity

A 50-ml aliquot from each of the 200-ml suspensions was filtered through a 0.2-μm Nuclepore polycarbonate membrane filter. The filters were transferred into scintillation vials containing 0.5 ml of 100% dimethylsulfoxide (DMSO) and were sonicated for 30 min using a sonic dismembrator. After sonication, 49.5 ml Milli-Q® reagent-grade water was added to each vial, producing a final DMSO concentration of 1.0%. The Microtox Screening Test[17] was then performed on these samples using a 15-min contact time.[18]

G. Contaminant Sorption by Suspended Particulates

Contaminant sorption studies were performed on water samples from the Yamaska River.[8] Two dyes, Acid Orange-60 and Basic Violet-1 (Figures 1 and 2) were used in this study (supplied by the Ecological and Toxicological Association of the Dye Stuff Manufacturing Industry, Washington, D.C.). Their octonol-water partition coefficients are given in Table 1. Portions of each of the 200-ml suspensions of suspended aggregate fractions were, separately, gently shaken with a 5-ppm solution of each of the dyes in order to establish to what degree they were bound to the various fractions. Samples from each dye mixture were centrifuged at 500 × *g* for 10 min, and the dye concentrations within the "dissolved" phase were determined spectro-photometrically using a

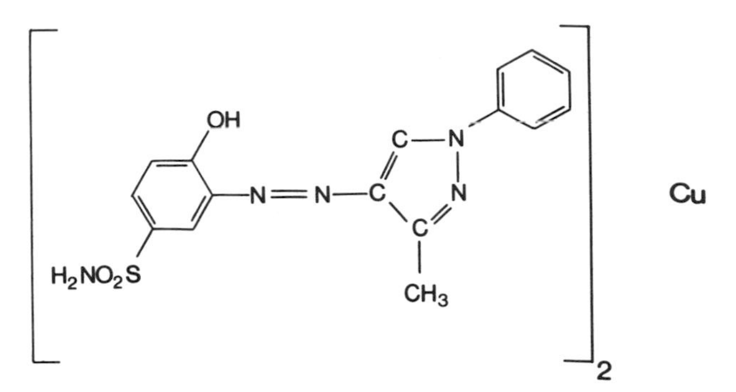

FIGURE 1. Structure of Acid Orange-60. (From Rao, S. S. et al., *Env. Toxicol. Water Qual.*, 7, 247, 1992. With permission of copyright owner© John Wiley & Sons.)

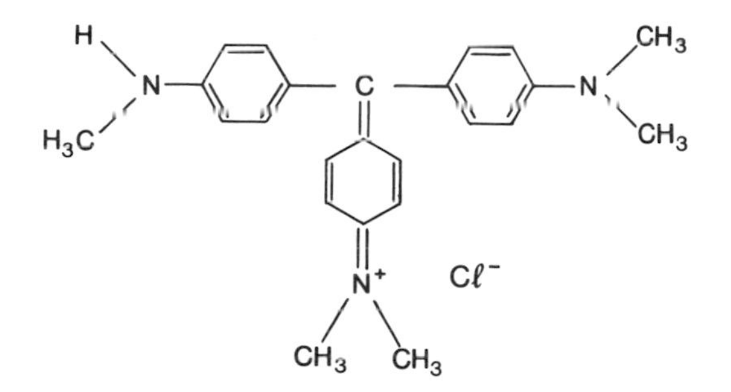

FIGURE 2. Structure of Basic Violet-1. (From Rao, S. S. et al., *Env. Toxicol. Water Qual.*, 7, 247, 1992. With permission of copyright owner© John Wiley & Sons.)

Table 1. Chemical Characteristics of the Dyes

	Chemical	
Characteristic	**Acid Orange-60**	**Basic Violet-1**
Chemical abstracts service number	30112-70-0	8004-87-3
Colour index number	18732	42535
Chemical formula	$C_{32}H_{30}N_{10}O_6S_2Cu$	$C_{24}H_{28}N_3Cl$
Molecular weight	778.33	393.96
Log of the octanol-water partition coefficient	0.66 +/– 0.24	–(0.17 +/- 0.05)

From Rao, S. S. et al., *Env. Toxicol. Water Qual.*, 7, 247, 1992. With permission of copyright owner© John Wiley & Sons.

wavelength of 560 nm for Basic Violet-1 and 273 nm for Acid Orange-60, at 24-hour intervals, for 96 h.

The sorption experiments were performed in the dark to exclude any photodegradation. Controls, without any suspended particulates, were employed as a check against chemical degradation and the possibility of adsorption of the dye onto the container walls. The contaminant adsorption results were calculated for each suspended particulate-size group and expressed in terms of the amount of dye (μg) removed (per milligram dry weight of particulate material) from the experimental units.

III. RESULTS AND DISCUSSION

Blachford and Day[1] indicated that large portions of transported fluvial particulates in Canadian rivers occur in the form of suspended materials in the 2- to 62-μm size range. Suspended particulates are usually composed of organic materials in association with inorganic materials such as clays and silts. The organic components of the suspended particulates are made up of bacteria and plant and animal materials of various origins, in various stages of decomposition. This provides a nutrient base for bacterial growth and enzyme production.

A. Suspended Particulate Distributions

From Tables 2, 3a, and 3b, it is evident that the 20 to 40-μm-sized suspended particulates are more prevalent or stable than are all of the other sized suspended particulates. This 20 to 40-μm-sized fraction accounts for approximately 30% of the suspended particulate material in the Yamaska River, 33% in the Athabasca River, 40% in the Thames River, and 30% in the St. Maurice River. The concentration of suspended particulates greater than 40 μm (40 to 88 μm) was approximately 26% in the Yamaska River, 50% in the Athabasca River, 47% in the Thames River, and 56% in the St. Maurice River.

B. Bacterial Content of Suspended Particulates

It is apparent from Table 3B that the 20 to 40-μm-suspended particulate size range generally contained more bacteria than did the other suspended particulate size ranges, as seen in three out of four rivers studied. In the Yamaska River, bacterial densities for the 20- to 40-μm fraction was 8.3×10^5/ml, while this fraction in the Athabasca River contained 1.4×10^8/ml, 7.2×10^8/ml in the Thames River, and 18×10^8/ml in the St. Maurice River. To our knowledge, this is the first documented evidence showing an association of bacterial populations with different size classes of suspended particulates.

C. Contaminant Adsorption by Suspended Particulates

This study was performed using suspended particulates from the Yamaska River.[8] The adsorption responses of two dyes, Acid Orange-60 and Basic Violet-1, for different particulate size classes are shown in Figure 3. With Acid Orange-60, an apparent equilibrium was reached in approximately 48 h with concentrations in the range of 5 to 45 μm of Acid Orange-60 per milligram dry weight of suspended particulate material. This dye was bound to particulates in decreasing concentrations relative to the following size fractions: (20 to 40) > (40 to 60) > (8 to 20) > (>88 μm). The logarithm of the apparent partition coefficient for this dye between the "dissolved" phase and the five size fractions (microgram dye per

Table 2. Concentration of Bacteria, Nutrients and Contaminants of Various Sized Suspended Particulates (Yamaska River, Quebec)

Size range (μm)	By vol. (%)	Bacteria (X10[5]/ml)	Parti. organic carbon (mg/l)	Parti. organic nitrogen (mg/l)	Microtox toxicity (EC_{50})
3–8	18.5 ± 6.1	3.5	ND	ND	ND
8–20	24.6 ± 1.7	3.7	5.06	0.39	44.90
20–40	30.4 ± 1.5	8.3	7.74	0.68	4.62+
40–64	14.0 ± 2.7	6.8	1.61	0.10	13.11
64–88	7.1 ± 3.2	3.6	0.75	0.04	90.89
>88	6.1 ± 2.4	ND	3.17	>0.002	6.52

Note: Size range was measured by a Malvern Particle Analyzer; +, most toxic; ND, no data.

From Rao, S. S. et al., *Env. Toxicol. Water Qual.*, 7, 247, 1992. With permission of copyright owner© John Wiley & Sons.

Table 3A. Suspended Particulate Distribution Pattern in Some Canadian Rivers (Dry Weight)

Size (mm)	Athabasca River (Alberta)		Thames River (Ontario)		St. Maurice River (Quebec)	
	(mg/l)	(%)	(mg/l)	(%)	(mg/l)	(%)
>88	4.4000	19.8000	0.0035	15.8000	1.7600	8.7000
60 – 88	4.5000	20.2000	0.0030	13.5000	4.6200	22.900
40 – 60	2.2000	9.9000	0.0040	18.0000	2.9000	14.3000
20 – 40	7.5000	33.7000	0.0089	40.2000	5.9400	29.5000
10 – 20	3.6000	16.3000	0.0027	12.2000	4.9200	24.4000
Total	**22.2000**		**0.0221**		**20.1400**	

Table 3B. Bacterial Content of the Different Suspended Particulate Sizes

Size (mm)	Athabasca River[1]		Thames River		St. Maurice River	
	(x10[8]/ml)	(%)	(x10[8]/ml)	(%)	(x10[8]/ml)	(%)
>88	5.4	17.7	1.8	12.2	6.1	13.8
60 – 88	7.1	23.2	2.3	15.6	4.2	9.5
40 – 60	9.3	30.5	2.4	16.3	5.8	13.2
20 – 40	1.4	4.6	7.2	48.9	18.0	41.0
10 – 20	7.3	23.9	1.0	6.8	9.8	22.3
Total	**30.5**		**14.7**		**43.9**	

[1]Average of three sites.

liter solution: microgram dye per kilogram suspended particulate material [dry weight]) varied from 3.6 to 4.1 between 24 and 96 h. It was estimated that, under the conditions of the experiment, about 20 to 30% of the dye was bound to particulate material >8 μm.

With Basic Violet-1, equilibrium was not reached even after 96 h, at which time the dye concentration in the suspended particulate material was in the range of 2 to 12 μm of Basic Violet-1 per milligram of suspended particulate

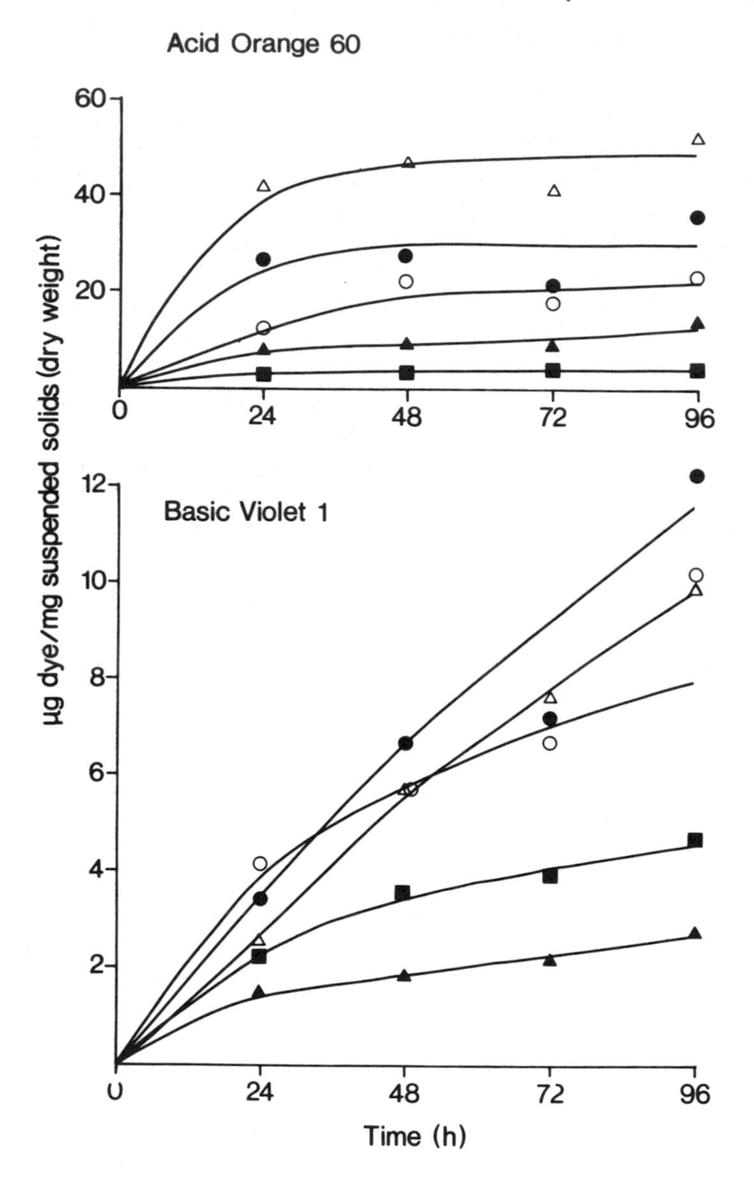

FIGURE 3. Rate of adsorption of Acid Orange-60 and Basic Violet-1 to different size fractions of suspended solids from the Yamaska River. ■, >88; ▲, 64 to 88; ●, 40 to 60; n, 20 to 40; ○, 8 to 20. From Rao, S. S. et al., *Env. Toxicol. Water Qual.,* 7, 247, 1992. With permission of copyright owner© John Wiley & Sons.)

material, dry weight. At 96 h, the concentration of Basic Violet-1 as a function of size fraction declined in the series (40 to 60) > (8 to 20) > (20 to 40) (>88) > (64 to 88 µm). It was estimated that, under these experimental conditions, at the end of 96 h, 5 to 10% of the dye was bound to particulate material >8 µm.[8]

D. Toxicity of Suspended Particulates

An assessment of the concentrations of the bioavailable toxicants on each fraction collected from the Yamaska River was carried out using the Microtox procedure.[17] The 20 to 40 μm-suspended particulate-sized fraction was found to contain the greatest concentration of toxicants.

The observation that the highest bacterial density and the greatest toxicant concentrations occurred within the same fraction is interesting in that this relationship had been observed in several different rivers. This suggests that either or both of the following hypotheses may be valid: (1) that bacteria are attracted to and therefore are able to concentrate toxicants, or (2) that the extracelluar polymeric substances secreted by bacteria may have an affinity for toxicant/contaminant molecules.[2-6] Either of these possibilities may account for the presence of higher toxicant concentrations in the 20 to 40 μm bacteria-rich fractions. Furthermore, the bacterial-excreted extracelluar polymeric substances may delay or nullify the impact of these toxicants on the bacterial cells (see Chapter 2).

The results of our studies suggest that the extent of contaminant binding to suspended particles is more likely influenced by bacterial biomass and their extracelluar polymeric substances, rather than by, as previously believed, physical-chemical forces solely acting on the surfaces of these particles. If the physical-chemical forces were the main controlling factors for particulate formation then the smallest sized suspended particulate fraction, (3 to 8 μm) in the Yamaska River should show the greatest binding efficiency. This, however, was not observed in our studies.

The greater concentration of bacteria in the predominant suspended particulate fraction (20 to 40 μm) is important, as it implies that the bulk of absorbed contaminants is transported in association with the predominant size fraction. Therefore, modelers working in the area of aquatic contaminant transport should consider focusing some of their attention on the presence of specific sized suspended particulates.

IV. CONCLUSIONS

The results of this study, based on observations from some Canadian rivers, indicate that suspended bacterial particulates are capable of adsorbing toxicants and nutrients, and, thus, are directly involved in the transport of environmental contaminants. Also, in support of these findings, it is known that bacteria alone or as part of a floating particulate mass can travel at least 22 km in small, slow-moving streams[18] and over 100 km in larger, faster-flowing rivers.[19]

Our studies have shown that different bacterial concentrations occur in different sized suspended particulates with maximum populations in the 20 to 40-μm-sized fraction. A study by Smith et al.[19] suggested that bacteria may not contribute significantly to the adsorption process because of their seemingly low proportion compared to other materials, such as clays, detritus, and humic substances, in both the suspended and bottom sediments. However, in our study, we have shown that a relationship exists between bacterial populations in suspended particulate material and contaminant/toxicant and nutrient concentrations. Thus, we believe that bacteria are essential for the formation and stabilization of suspended particulates, and that bacteria likely play a key role in the adsorption and eventual transport of contaminants/toxicants in riverine systems.

It would be interesting to investigate whether the association of high organic content and toxicity with the predominant size fraction (20 to 40 μm) of suspended particulates is a natural phenomenon found in all rivers. If this can be substantiated, then this would confirm that bacterial biomass, toxicant, and nutrient loads are related to specific suspended particulate sizes. This could have significant implications in the development of contaminant-transport models.

REFERENCES

1. Blachford, D. P and Day, T. J., Sediment Water Quality Assessments: Opportunities for Integrating Water Quality and Water Resources Branches' Activities Sediment Survey Sec IWD-HQ-WRB-SS-88-2 Environment Canada, 1988, 1.
2. Fletcher, M. and Floodgate, G. D., An electron microscope demonstration of an acidic polysaccharide involved in adhesion of a marine bacterium to solid surface, *J. Gen. Microbiol.*, 74, 325, 1973.
3. Marshall, K. C. and Cruickshank, R. H., Cell surface hydrophobicity and orientation of certain bacteria at interphases, *Arch. Microbiol.*, 91, 29, 1973.
4. Marshall, K. C., Theory and practice in bacterial adhesion process. Trends in microbiological theory, in Proc. IV Int. Symp. Microbial Ecology, Yugoslavia, 1986, 112.
5. Beveridge, T. J., The bacterial surface: general considerations towards design and function, *Can. J. Microbiol.* 34, 363, 1989.
6. Beveridge, T. J., Role of cellular design in bacterial metal accumulation and mineralization, *Annu. Rev. Microbiol.*, 34, 147, 1989.
7. Karickhoff, S. W., Semi-empirical estimation of sorption of hydrophobic pollutants on natural sediments and soils. *Chemosphere,* 10, 833, 1981.
8. Rao, S. S., Maguire, R. J., Krishnappen, B. G., Weng, J. H., and Ongley, E. D., Effect of suspended aggregate sizes on the adsorption of water soluble dyes in an aquatic environment, *Environ. Toxicol. Water Qual.,* 7, 247, 1992.
9. Tate, D. M., Economic and Financial Aspects of Waste Water Treatment in the Yamaska River Basin, Quebec, Solid Sciences Service, No. 3, Inland Water's Directorate, Water Planning and Management Branch, Environment Canada, Ottawa, Ontario, 1972, 1.

10. Dutka, B. J., Kwan, K. K., and Rao, S. S., An Ecotoxicological — Microbiological Study of the Yamaska River, National Water Research Institute Contribution #89–147, Burlington, Ontario, Canada, 1989.
11. Bale, A. J. and Morris, A. W., *In situ* measurement of particle size in estuarine waters, *Estur. Coastal Shelf Sci.*, 24, 253, 1987.
12. Droppo, I. G. and Ongley, E. D., Flocculation of suspended solids in Southern Ontario rivers, in: *Sediment and Environment,* Hadley, R. J. and Ongley, E. D., Eds., Int. Assoc. Hydrologic Sciences, 3rd Scientific Assembly, Baltimore, Maryland, May 10-19, 1989, IAHS No. 1984.
13. Rao S. S. and Kwan, K. K., Method for measuring toxicity of suspended particulates in waters, *Toxicity Assessment: Int. J., 5*, 91, 1990
14. Marxsen, J., Evaluation of the importance of bacteria in the carbon flow of a small open grassland stream, the Breitenbach, *Arch. Microbiol.*, 11(3), 339, 1988.
15. Rao. S. S., Jurkovic, A. A., and Dutka, B. J., Some factors influencing the enumeration of metabolizing aquatic bacteria, *J. Test* (ASTM), 12, 56, 1984.
16. Water Quality Branch, Analytical Methods Manual, 1979. Inland Water Directorate, Environment Canada, Ottawa, K1A 0E7, 1979.
17. Beckman Instrument, Inc., Beckman Microtox System Operating Manual, NO. 015-555879, 1982.
18. Dutka, B. J. and Kwan, K. K., Application of four bacterial screening procedures to assess changes in the toxicity of chemicals in mixtures, *Environ. Pollut. Ser. A.*, 29, 125, 1982.
19. Smith, J. R., Mabey, W. R., Bohonos, N., Holt, B. R., Lee, S. S., Chou, T. W., Bomberger, D. C., and Mill, T., Environmental pathways of selected chemicals in freshwater systems. I Background and experimental procedures. U.S. Environmental Protection Agency Report EPA-600/7-77-113, Washington, D.C., October, 1977.
20. Geldreich, E. E., U.S. Environmental Protection Agency, Washington, D.C., personal communication.

CHAPTER 7

Organic Flocs in Surface Waters: Their Native State and Aggregation Behavior in Relation to Contaminant Dispersion

Gary G. Leppard

TABLE OF CONTENTS

0-87371-678-7/93/$0.00+$.50

I. INTRODUCTION TO AQUATIC ORGANIC FLOCS

A. Definitions and Context for Research

An aquatic organic floc is defined here as a suspended particulate with two properties: (1) it is rich in organic molecules and materials; (2) it is derived by an aggregation process in an aquatic ecosystem. The term "particulate" includes the native colloids whose least dimension falls between 1.0 and 0.001 μm.[1,2] It also includes organic-rich organo-mineral aggregates and submicron organisms, both living and disintegrating (especially bacteria, algae, and viruses).[1,3] Until 1988,[1] native heterogeneous suspended colloid systems had only occasionally been subjected, as flocs per se,[4,5] to structural analysis by the high resolution of transmission electron microscopy (TEM). For several decades, however, there has been developing a literature on individual floc components,[1-3,6,7] including the refractory cell parts of relatively large planktonic organisms.[1] Principal among the components are fibrillar extracellular polymeric substances, humic acids, fulvic acids, bacterial envelope fragments, algal cell-wall fragments, incomplete viruses, and organic coatings on minerals. There is also extant a related literature on the structure of complex biofilms.[8-10] As the understanding of floc components and their associations grows, the interest in the bulk composition of flocs is being replaced by an interest in specific floc types and their intrinsic properties as materials.[3]

The specific aggregation processes for floc formation are variously termed flocculation, coagulation, agglomeration, agglutination, or, simply, aggregation. The use of these terms depends on the scientific specialists involved and can refer to differences in mechanism.[11] In this brief review, the general aggregation processes under consideration will all be referred to as aggregation, except when a cited research makes specific reference to either flocculation or coagulation, both of which will be treated as similar in mechanism.[11,12] As a rule, for aquatic scientists, flocculation is aggregation due to polymers, and coagulation is aggregation due to electrolytes.[13] Polymer bridging allows small flocs to aggregate into macroflocs, as is described below.

Organic-rich flocs are produced from smaller particulates (which can even be as small as molecules of moderately high molecular weight)[1,2] by processes which usually involve some form of physical or chemical destabilization and a step in which the particulates collide.[11,13,14] In aggregation studies, the term "stability" is not used in the thermodynamic sense; a stable colloidal system is one which is slow in changing its state of dispersion during a period of observation. The common modes of destabilization (or change in state of dispersion) and their characteristics have been known and documented for decades;[12,15] this is true also for many sizing considerations.[16]

For flocs in the colloidal size range, recently applied scaling concepts[17] are improving our understanding of the structure of aggregates and the kinetics of their formation. As a consequence, two distinct limiting regimes of irreversible colloid aggregation have been identified, diffusion-limited and reaction-limited aggregation, with growth extending into the true particle (>1 μm) size range.[18] These regimes are differentiable by TEM,[18] as are details of substructure (e.g., porosity, surface-to-center heterogeneity) for a given growing floc.[3] For colloidal flocs (and their aggregates up to 10-μm diameter) terminal gravitational settling will likely be less than 10^{-2} cm s^{-1}.[16]

It has become evident that many individual components of an organic-rich aggregate can be visualized by TEM, at 0.001-μm resolution, **as individual entities**[1,19] taking up specific positional arrangements within a given floc.[2,3] The observation that many components appear to be sticking to each other side-by-side, as opposed to flowing into each other, is of considerable utility.[3] Because many identifiable classes[1] of colloids within a growing floc do not change size or shape as the floc size increases, one can thus study the relative contributions of each identifiable class to aggregation. Colloid identification begins by a consideration of size and shape and native electron-opacity. Then, accessory techniques are brought into the characterization as needed. These have been reviewed recently[1] and include microchemistry, cytochemistry, energy-dispersive spectroscopy and electron diffraction. Thus, TEM, in conjunction with: (1) an improved physical and chemical information base,[11,18,20-22] (2) more sophisticated analytical accessories,[1] and (3) more versatile multimethod approaches,[1,19] is featured in this brief review as a principal means of investigating aquatic organic flocs.

Central to its descriptive role, there is a major challenge for TEM. This challenge is to help integrate a molecular understanding of key aquatic chemical reactions with the specific characteristics of native flocs, especially those relevant to contaminant dispersion phenomena (the temporal and spatial re-distribution of dissolved and suspended contaminants). Currently, TEM has the potential to inform us profoundly about the disposition of molecules and materials within heterogeneous native flocs, and do it realistically with minimally perturbed specimens.[19] Used in conjunction with a rapidly improving information base in the aquatic sciences, TEM should facilitate analyses of floc behavior in aquatic ecosystems, including the coupling of adsorption and sedimentation.

B. General Significance for Contaminant Dispersion

Organic-rich flocs and the organics contributing to floc formation (colloids and surface-active molecules) are increasingly recognized for their important roles in contaminant binding.[23-26] General treatments (mathematical, physical, chemical) of colloid aggregation and behavior are available with reference to floc growth and concomitant changes in contaminant status (soluble vs. suspended).[27-30] Engineered systems have been under study for some time with respect to analyzing relations between aggregate growth, particle behavior, sedimentation, and contaminant removal from water.[31,32] As special cases of organic flocs, one can consider biofilms,[8,10,33] surface films,[26,34] and coatings for informative features;[7,35-37] with the exception of coatings, such special cases will be considered only briefly here.

The major classes of native organic molecules which enter into aggregation processes in surface waters are described in detail elsewhere.[34,38] Prominent among these are (1) macromolecules and high polymers, well exemplified by polysaccharides and polysaccharide-protein associations;[10,24,39] (2) some of the humic substances,[40-42] as exemplified best by the complex versions found in nature, as opposed to the degraded humics derived from degradative separation procedures. Such organics can readily interact with contaminants of many kinds to alter their activity in water. In this vein, it is increasingly evident that descriptions of the "speciation" of hydrophobic contaminants[43] in natural waters should include not only dissolved and true particle fractions, but also an important separate fraction of "contaminants sorbed to macromolecules and colloidal particulates which are essentially nonsettling".[43]

There is growing evidence of a direct coupling[27] of contaminant adsorption (by colloids) to aggregation processes which lead to the formation of contaminant-bearing settling flocs (and ultimately to contaminant transport and/or burial). Such research is being followed with great interest, while information gaps constraining its progress are being placed under scrutiny. One gap is the need for more knowledge on quantitatively important classes of aggregates. These are distinctive materials with their own intrinsic properties and should be studied as such in a state as close to the native one as is possible;[3] a colloid system is not an unstructured bag of molecules. This goal, in turn, requires the development of an ultrastructural literature on unperturbed, native, aquatic colloids and flocs.[1] In cases of specific contaminant/particulate associations in surface waters, the prediction of residence times for the contaminants is often not possible without better information on the dynamics of the relevant particulates.[27,44] When the particulates of concern are flocs composed of colloid systems, then an understanding of the dynamics requires an understanding of floc substructure. This understanding is best achieved by the TEM characterization of minimally perturbed flocs. TEM examinations can yield helpful information on shape, size, size distribution, porosity, and elemental composition, as well as revealing the presence of polymer bridges, microcrystalline regions, and selected types of macromolecules. For current studies on contaminant partitioning into growing aggregates, one can use the crude aggregate classification criteria now extant[1,3] (e.g., fibril-rich, iron-rich, clay-rich, virus-rich, etc.).

C. Structure and Dynamics: Their Connection at the Macromolecular Level

For biological macromolecules, structure and function meet near the 0.001 μm practical resolution achievable by TEM. This dimension is at the lower limit for colloid diameter, thus permitting the visualization of colloids throughout their entire size range. It is also below the upper size limit for macromolecules; thus, large molecules can be visualized in an electron micrograph, although only as dots or lines by most conventional image-processing techniques. The localization of classes of structural molecules (such as those of membranes) and the localization of enzyme activity to specific sites within colloid systems (usually cell compartments) are well-developed arts within ultrastructural biology. To a limited extent, the TEM-related technology is transferable to analyses of the nonliving organic components of flocs,[1,2] and, to a great extent, one can use TEM to characterize the molecular architecture of the living components, especially bacteria.[6,45,46] Relating molecular "activity" and "speciation" (in the form of structure-function correlations) to TEM images has, in fact, been serving science well for many decades,[47] despite an often large trial-and-error aspect. The relevance of this technology, literature and experience to organic floc investigations is that they indicate a real potential for studying phenomena such as:

1. The internalization or occlusion of reactive materials within a single colloid
2. The extent of coverage of particulates by coatings
3. The porosity or various differing porosities within a colloid system
4. Structural and enzymatic gradients traversing a suspended macrofloc or a biofilm
5. Three-dimensional relationships between different identifiable materials within a submicron floc
6. Effects of dissolution on the release of internal materials by a floc
7. The nature of aggregation of submicron aggregates, whether reaction-limited or diffusion-limited
8. The relative amounts of inorganic and organic colloids within a given class of floc, done on a per-floc basis rather than on a bulk-analysis basis.

Knowledge of such phenomena can improve one's interpretation of dynamics. When details of ultrastructure can be obtained in sufficient depth, with regard to a given floc or colloid system in water, one can generate revealing hypotheses about dynamics and test them with experiments. This approach has always worked well for cell biologists[47] and is currently providing insight into the dynamics of iron-phosphate colloids and their aggregation in lake water.[48]

D. Biological Roles in Flocculation Processes

To their aquatic milieu, many kinds of biota can supply a variety of molecules and colloidal materials which facilitate aggregation,[10,24] including true flocculation. Many bacteria,[24] prokaryote algae,[39,49] eukaryote algae[24,49] and higher plants (their roots)[1,50] secrete fibrillar extracellular polymeric substances[10] (or fibrils)[39] which can detach from cells to enter the bulk water phase. While suspended, fibrils behave as adhesives, sometimes being visualized by TEM as polymer bridges[4,5] between colloids in a floc. Many microbes use fibrils in a direct manner to attach themselves to surfaces;[24] they literally glue themselves into position by means of an essentially two-step process in which copious fibril secretion is preceded by a weak attachment phase (the initial reversible sorption step).[51] Additional major contributions by biota to aggregation include: (1) a microbial contribution to the processing of refractory organics into humic substances which can aggregate and (2) a zooplankton contribution to agglomeration by packaging small digested aggregates into relatively large fecal pellets. These latter two topics will not be pursued further here.

A detailed discussion of biological roles in flocculation processes is beyond the scope of this review. Bacterial roles, however, having been investigated in great depth with respect to biochemistry and dynamics, will be considered briefly. There are literature reviews of bacterial products and structure relevant to adhesion and aggregation phenomena[6-8,10,24,33,39,45,46,49]. The analysis of bacterial contributions to the growth and differentiation of a suspended floc is complicated by population diversity and the differing physiological requirements for the growth of each species. Spatial heterogeneity in bacterial ecosystems is an established phenomenon.[52] It is a direct consequence of several biological species responding differently to a variety of physical and chemical gradients. This leads, in turn, to the establishment of microniches in which one can observe, by TEM, a variety of extracellular macromolecular species, differing from one microniche to another.[8] To complicate the situation further (and to make it more challenging), one finds the following additional phenomena, many of which require TEM for analysis:

1. A free cell may behave (to an experimenter making counts) temporarily as an adherent one without actually adhering.[33]
2. Several morphologically different exopolymers (fibrils) may be produced by a given bacterium, with not all types used to form polymer bridges to a surface.[24]
3. Several different exopolymers may be produced in a sequence.[24]
4. Exopolymers produced during one phase of population growth may be different from exopolymers produced during another phase.[24]
5. Some adherent bacteria are capable of reversible attachment, sometimes by means of the secretion of a hydrophilic exopolymer which counteracts the initial adhesive (a hydrophobic exopolymer).[24]
6. The nature of the exopolymer can be dictated by the nutrient regime.[24,49]

7. The amount of exopolymer secreted can be dictated by the nutrient regime.[24,39]
8. Some bacteria produce exopolymers which are temporary; they lose their association with a surface quickly relative to the life of a cell.[24]
9. Some bacteria are specialized for attachment to exuding surfaces only (plant and animal surfaces), being able to attach only at specific positions using distinctive, specialized extracellular structures.[24]
10. An exopolymer can adsorb and concentrate many substances from water to change its own chemical composition.[24]
11. A given bacterium can have separate mechanisms for adhesion to hydrophilic and hydrophobic substrata.[53]
12. Bacteria can mutate to lose their capacity to generate exopolymers and, consequently, their ability to aggregate.[54]
13. Exopolymers can be so abundant as free fibrils on an episodic basis (7 mg/L)[55] that they are likely to form coatings on the surfaces of suspended flocs regardless of the direct bacterial contribution of fibrils.[36]
14. Fibrillar exopolymer gels in fibril-rich flocs can drastically change their three-dimensional disposition when the floc is subjected to harsh dehydration[2,19] by analysts prior to analysis.

This list is not presented to discourage research; rather, it is meant to encourage a high level of planning prior to action.

II. SELECTED CASE STUDIES OF FLOC/CONTAMINANT ASSOCIATIONS

A. An Early Model System Approach

Several decades ago, a team of microbial and cell biologists combined TEM with pollution studies in a manner so revealing that it has guided aquatic scientists right into the present. In essence, they (1) took quantitatively important floc-forming bacteria into the laboratory, (2) analyzed the chemical structure and (3) physical structure of the secreted floc-matrix materials, (4) investigated the uptake of contaminants by the matrix materials, and then (5) used their new-found knowledge to sample and analyze natural waters for sedimenting flocs implicated in contaminant transport. Their results are treated briefly below.

Zoogloea-producing bacteria (those producing a gelatinous matrix leading to floc formation) are common in fresh surface waters. They can grow in the presence of high concentrations of many contaminants[56] and are readily isolated from oxidative waste-treatment systems.[56] Initial results showed the floc matrix material to be polysaccharide in nature[57] and organized into discrete colloidal fibrils.[57] The fibrils were analyzed by a multi-method approach to rule out the possibility of their fibrillar aspect being an artifact of TEM preparatory techniques.[57] Analyses of the fibril-bacterial associations clearly suggested that entanglement of cells among fibrils and adsorption of cells to fibrils were mechanisms of floc formation.[58]

Studies of metal accumulation by flocs (cells plus fibrils) showed a great accumulation of copper and cobalt (on the basis of floc weight).[56] After considerable further success with metal-binding studies, they turned their attention to organic contaminants. From a collection of floc-forming bacteria isolated from Lake Erie, two kinds were examined for their ability to accumulate aldrin (a chlorinated hydrocarbon pesticide) from solution.[59] These flocs proved to be excellent for the adsorption of aldrin from their aquatic milieu, giving a 625 × concentration factor within 20 min,[59] with a recovery value averaging 88%.[59,60] While suspended, the floc acted as an adsorbent; through settling, it transported the aldrin out of the water column to the bottom of its container. Subsequently, the team moved closer to nature, demonstrating that floc-forming bacteria in Lake Erie do tend to play a role in the purification of the water column. While growing, Lake Erie flocs adsorb and concentrate aldrin from colloidal dispersion. Upon settling, they formed conglomerate[59] flocs in the sediment (with diatoms, detritus and inorganic particles, as revealed by TEM), and the conglomerates tended to retain the aldrin.[60]

B. Organic Contaminants

The transport of organic contaminants on organic-rich flocs occurs in surface waters; there is considerable evidence to indicate that it is an important means of contaminant dispersion[23,59-61] and part of a natural purification mechanism. The theoretical basis for treating this natural mechanism experimentally is improving rapidly in relation to the sequence: (1) sorption, (2) aggregation, (3) sedimentation and (4) burial.[30] Many specific interactions relating to this sequence have been reviewed recently.[23] The greatest information gap resides in the paucity of detailed characterizations of the floc types involved.

The greatest interest in flocs as agents of transport (and in the partitioning involved) currently revolves around colloids and colloid systems.[23,43,62,63] Colloid systems include macromolecular gel networks where sorption can be viewed as a dissolution into an organic polymer matrix,[64] a situation which can lead to poor recovery of a contaminant from nature for quantification by analytical laboratories.[64]

Contaminant transport by organic-rich colloids appears to be important for atrazine,[61,63] azobenzene,[65] benzidine,[65] endrin,[66] lindane,[66] linuron,[61] mirex,[67] various polycyclic aromatic hydrocarbons,[68] and toluidine.[65] Assuming colloidal components of macroflocs to have been active in sorption during floc growth, one might add aldrin [59,60] to the list. A consideration of resuspendable pore-water colloids would certainly produce more examples, such as polychlorinated biphenyls.[69,70]

C. Metals

Surface waters have such a great diversity of biota with such a variety of nutritional needs that, barring a pollution event, there are difficulties with considering metals as toxic contaminants in other than a vague sense. A given metal is toxic for a given organism in an ecosystem if it damages life functions, but is nontoxic otherwise. This point is emphasized to avoid creating misleading linkages between bioaccumulation, bioconcentration factor, toxicology, and (purely physico-chemical) partitioning for relatively unstudied floc-forming organisms and their fibrillar products in complex, natural waters. Many "so-called toxic" metals are, in fact, essential nutrients,[71] being toxic only at concentrations above their natural nutritional levels[71] (that is to say, at "stressful concentrations"). High concentrations of metals (e.g., Cu, Co, Pb, Zn, Cd, Mn, Fe, Ni, Sr, Ag, and others) often accumulate in flocs rich in extracellular polymeric substances,[24] especially fibrils composed largely of acid polysaccharides.[24] Much of this binding can be attributed to the avoidance of stressful metal levels in the aquatic milieu by organisms having a capacity to immobilize metal ions on a surface which is external to the metabolic machinery of living cells (a surface represented by the extracellular fibrils). The affinities of fibrillar exopolymers for metals vary, depending on the specific metal involved, but many empirical measurements indicate high weight-specific binding capacities.[24] Metal binding to the molecules of exopolymer is strongly influenced by pH,[24] but not all binding appears due to adsorption. Precipitation reactions followed by physical entrapment of a precipitate in a fibril matrix[32] can effectively remove metals, diffusing in water, from contact with the uptake sites on the cell membrane of a floc-forming organism. Between adsorption and precipitation reactions, metals can bind in such considerable amounts to fibril-rich flocs that they become converted into rapidly-sinking particles.[72]

The literature on the binding of metals to fibrils is extensive and scattered among many disciplines. It describes many case studies based on both metal species and biological species. The biological species are represented strongly by Gram-negative bacteria and small algae, both prokaryotic and eukaryotic. Guides to the various specialized topics (as they have developed over several decades) are found by consulting several references[7,10, 24, 32, 38, 39, 50, 56]. The chemistry of the exopolymer molecules, especially from bacterial sources,[73] is now under detailed scrutiny. There are related publications on anomalous metal behavior in surface waters.[27,44] These include the phenomenon of "colloidal pumping",[27] which is the transfer of sorbed metal species from the colloid pool to larger (sedimenting) aggregates in natural waters.

While the fibrils are important metal-binding components of many organic-rich flocs, there are others of acknowledged significance: (1) organic coatings and (2) the cell walls of bacteria and algae. Each of these components may also contain fibrils or be associated with fibrils.[36] It has been accepted for many years that surface active molecules in aquatic ecosystems will spontaneously form coatings around immersed particles.[35,37] Humic substances are implicated in such coatings, including colloidal humic substances[40-42] which can also aggregate into fibrous networks and even into large colloids approaching the true particle size range.[41] The coatings, in addition to binding metals[74,75] and modulating aggregation,[37] can affect metal transport by altering the settling velocity of particulates[76] by means additional to increasing the aggregate size. The importance of coatings to biogeochemical processes is such that it has become essential to characterize them better, along with their spatial associations in heterogeneous metal-scavenging aggregates. To this end, new analytical approaches are being developed to determine their nature and coverage on a particle-by-particle basis. These approaches include techniques based on TEM (in association with energy-dispersive spectroscopy, electron diffraction, and selective staining)[1] and on analytical flow cytometry (in conjunction with selective staining).[77]

When one considers bacterial aggregates, with or without extensive fibril bridging, one must consider also the metal-selective cell envelope of individual bacteria (membrane plus wall plus extra wall layers). The bacterial envelope interacts with metals in complex and highly specific ways. The ultrastructure, chemistry, and reactivity of the surface layers of the envelope have been reviewed recently,[6] as have envelope roles in bacterial metal accumulation and mineralization.[45] The thickness of these layers is measured in thousandths of a micrometre, but through TEM analyses of selectively stained (by metals) walls, one can visualize even the discontinuity of charge on the surface of a single bacterium[78] at its cell wall-aquatic milieu interface. The significance of biologically-directed envelope activities, with regard to floc behavior in relation to contaminant transport, is that envelopes provide nucleating interfaces for removing metals from dilute solution. The formation of facilitated metal precipitates at such interfaces leads to an increase in density for the organic floc (when it is considered as a sedimenting unit possessing an increasing proportion of metal); thus, an active floc can become a transporter of metals by coupling metal binding with sedimentation. In experimental systems,[79] preferential iron sorption onto the bacterial wall component of an envelope can lead to the formation of macroscopic flocs from suspended colloids. The flocs in one system have been shown to be composed of walls glued together by iron compounds.[79] In a related experiment, the sorption of iron onto walls was found to enhance the uptake of other metals.[79]

D. Merits and Flaws of Earlier Studies

Earlier studies of natural flocs and aggregation processes in surface waters were hampered by an insufficient understanding of colloid instability artifacts, a tendency for chemists and biologists to avoid communication with each other and a reluctant appreciation of the quantitative importance of aquatic colloids. The fact that contaminant associations with organics (cells, materials, molecules) are traditionally related to a filtration operation (and that this operation defines the association without reference to colloids) has caused considerable confusion. This confusion, however, should cease as a result of advances in our understanding of (1) the mechanics of filter fractionation operations[20,80] and (2) an improved capacity to detect, assess, and minimize colloid instability artifact.[1,3] Despite its flawed history, the field has advanced, and a solid foundation for future research has been established. Flaws which generated misleading literature will be discussed in Section V.

The merits of earlier studies reside in many of the works described previously herein: (1) the identification of major floc constituents; (2) the creation of sound theoretical bases for designing experiments; (3) the development of concepts to explain apparently anomalous colloidal behavior in natural systems; and (4) the establishment of an information base for linking the binding of contaminants to aggregation phenomena and subsequent contaminant transport. The useful portion of the pre-1990 information base is identifiable and sufficiently advanced to permit one to attempt realistic structural analyses of native, aquatic flocs, including their physically unstable colloidal components. Improved analyses of dynamics should follow.

The flaws of earlier studies have led to information gaps, some of which can now be addressed by the use of TEM, as a result of recent technological advances.[1,6,19,45] Structure-based flaws have entered the literature, mainly as a result of inadequate sampling, sample handling, and sample storage (including preparation for TEM analysis),[1,81] as well as too great a reliance on simplified model systems. In fairness to past researchers, it must be noted that much of what they did was appropriate for the information needs of the moment. The consequence of their efforts, however, is a need to create a **new literature**, one providing detailed characterizations of **minimally-perturbed native** flocs and the **hydrated components** within them, especially those which had been implicated previously in contaminant dispersion. To understand dispersion in a complex and varying natural setting, it is clear that one must understand the physical behavior and biogeochemical characteristics of the dispersing agents.

Floc matrix materials and extracellular polymeric substances have received considerable attention as modulators of water chemistry, especially fibrillar ones,[3,4,7-10,24,36,39,49,57,58] but the descriptive phase remains incomplete for them. They require a greater consideration in analyses of realistic (minimally-perturbed) three-dimensional reconstructions of hydrated floc matrices. The technology is now available[1,3] to analyze them in relation to secretory surfaces, extracellular enzyme positioning, biomineralization activity, and tendencies to form natural associations in a selective manner with common aquatic particulates. Specific information needs are treated well in Decho.[24]

III. STATE-OF-THE-ART CHARACTERIZATION OF NATIVE FLOCS

A. What is Needed? — How to Achieve It!

What is needed most is to better bridge the gap between (1) those physical and chemical techniques which permit atomic scale resolution and (2) the technology of correlative microscopy.[3] An improved bridging will allow one to relate floc structure and development more precisely to adhesion phenomena[8,24] and contaminant partitioning,[24,27,28,30,59,60] as well as to relate all of these phenomena in nature to specific environmental and economic problems.

Figure 1 illustrates, at atomic resolution, a physico-chemical representation of a generalized aquatic particle.[82] Additional structural considerations, at a mesoscopic[83] level and relating to hydrated organic particles, are readily found in the literature.[84,85] Figures 2 and 3 show two kinds of organic flocs (and their particulate subcomponents) prepared by state-of-the-art, minimally perturbing TEM techniques.[1,19] Figure 2 reveals a native floc rich in bacteria and fibrils,[86] whereas Figure 3 shows a floc composed of large molecules which had been isolated from lake water and then concentrated in solution so as to provoke floc formation.[87] With floc preparations such as these (when analyzed intensively at high primary magnifications), one can achieve more realistic characterizations (than was permitted previously near the practical resolution limit of TEM) if one employs the general strategy outlined below.

The general strategy must employ a **minimal perturbation approach** which includes TEM revelation of floc components in their hydrated state. The technical means to realize this strategy, as well as the scientific basis for justifying increased research resources, have been reviewed recently,[1-3] including an elegantly brief treatment on the benefits of *in situ* analyses of the microbes within natural aggregates.[88] Coupled with the minimal perturbation approach, one must strive for **multi-method analyses** so that each environmentally significant result receives support from two or more complementary but independent, techniques.[1] The multi-method analyses should include, to the greatest extent feasible, conventional wet chemistry carried out directly in the field,[1] in conjunction with field-based TEM preparations.[19]

B. Minimal Perturbation Approach

The less one perturbs the sample prior to **careful examination**,[3,89] the more readily one can interpret the finest details of ultrastructure which relate morphology to chemistry. Minimizing perturbation must be done at all stages of sample handling, not just at those stages involved directly with preparation for electron microscopy. Where chemical entities in companion studies are sensitive to artifactual change during sampling/processing, one must organize so as to minimally perturb the relevant molecules (e.g., changes in oxidation state); otherwise one risks misinterpreting chemistry in relation to morphology. To

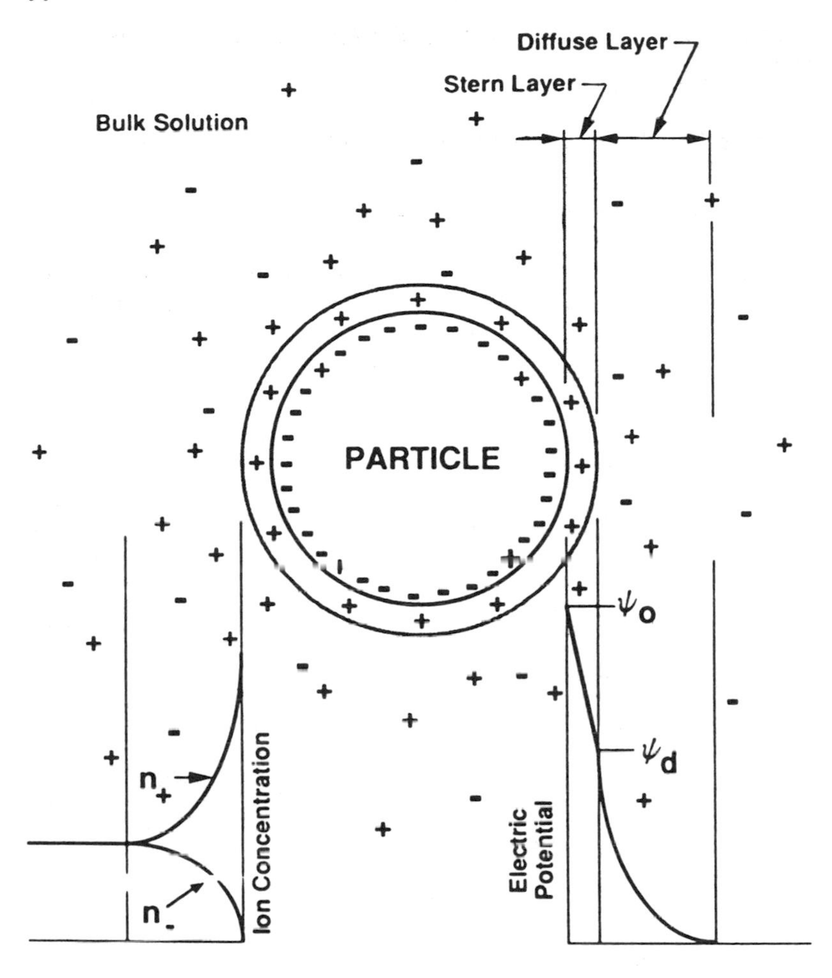

FIGURE 1. A physico-chemical representation of an aquatic particulate at atomic resolution. It shows, schematically, an electrical double layer surrounding a solid particle in aqueous "solution", where ψ_o and ψ_d are surface and diffuse layer potentials, respectively, and n_+ and n_- are cation and anion concentrations. (Adapted by O'Melia, C. R., *Aquatic Chemical Kinetics — Reaction Rates of Processes in Natural Waters,* Stumm, W., Ed., John Wiley & Sons, New York, 1990, 47; as adapted from Hirtzel, C. S. and Rajagopalan, R., *Colloidal Phenomena: Advanced Topics,* Noyes Publishers, Park Ridge, NJ, 1985, 473. With permission.)

the greatest extent feasible, it is wise to coordinate one's minimal perturbation efforts at all levels of one's research (limnological, chemical, physical and microbiological). When the construction of a three-dimensional model of a specific floc type is the research goal, one should consider that different components of the floc may be subject to artifactual alteration by a given procedure to differing extents (e.g., nonextractable crystals and rigid materials may be insensitive to artifact relative to fibrous floc matrices). The major

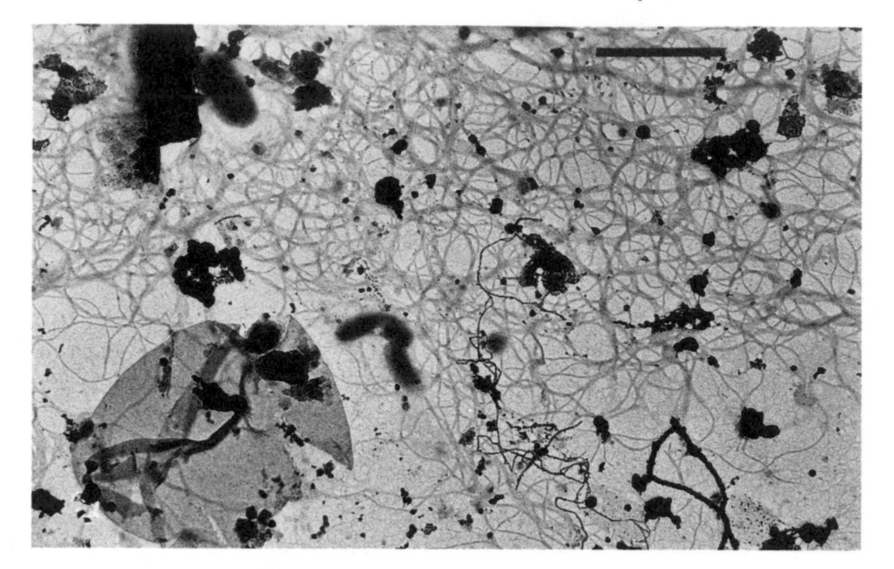

FIGURE 2. A heterogeneous organic-rich floc taken from Lake Bret (near Lausanne, Vaud, Switzerland). This floc was isolated from lake water in a minimally perturbing manner using the Nanoplast film technique of Perret et al.[19] The floc is presented visually as an unstained whole mount[1] which had been embedded at the sampling site directly and immediately in a water miscible resin. There are no artifact contributions at all from chemical fixatives, dehydrating solvents or the sample handling associated with standard TEM preparations for embedding in a hydrophobic resin. Note the unusually small bacteria and the fibrillar extracellular polymeric substances which intertwine to give rise to a porous floc matrix. Bar = 1.0 μm. (From Leppard, G. G., Belzile, N., Perret, D., Filella, M., and Buffle, J., unpublished results.)

causes of artifact among the "amorphous" organic components of organic flocs processed by **standard**[90] electron microscope techniques are

1. Harsh dehydration[19] (e.g., air drying or bulk freeze-drying) which can cause severe shrinkage and do it unevenly, so as to provoke a positional reorientation of floc components
2. Extraction[2] (e.g., by washes and dehydrating solvents) which can result in a selective loss of some components and a spatial redistribution of others
3. The use of reagents in chemical preservatives[90] which affect the activity of some floc components (e.g., enzymes) during the processing period, including activities which one might wish to measure *in situ* in microscope preparations by cytochemical techniques[1]

For high-resolution ultrastructural analyses, it is obvious that even small problems of artifact, if left unresolved, will lead to difficulties in interpretation of TEM images. The problem of artifact can appear in the field or in the laboratory, prior to analysis or during analysis, during the sampling or while processing the samples; even standard methods to fractionate, purify, or concentrate the sample can introduce artifact.[1-3,19,20,27,41,48,80] Despite the difficulties

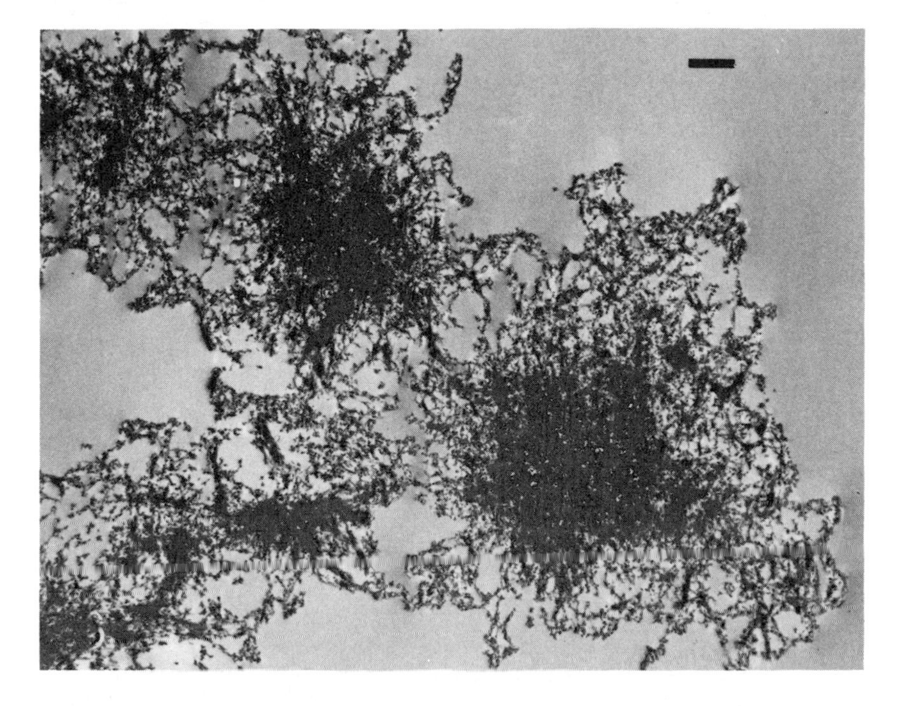

Figure 3. Organic flocs composed of high-molecular-weight natural compounds isolated from epilimnetic waters of Bay Lake (northeast of Peterborough, Ontario, Canada). A water fraction (whose organic materials were smaller than 0.2 µm and whose molecules were larger than 30,000 Da) was provoked to aggregate by increasing the solution concentration factor by several hundred times the lacustrine concentration.[2] Growing flocs are presented in section view (0.05-µm section thickness) after embedding of the fully hydrated flocs in a water miscible resin.[2] There are no artifact contributions at all from chemical fixatives, heavy-metal stains or dehydrating solvents; storage of the flocs was minimized.[1] Note that the size of individual provoked flocs is nearly two orders of magnitude larger than the largest materials which took part in the flocculation process. Bar = 1.0 µm. (From Leppard, G. G., Burnison, B. K., and Carey, J. H., unpublished results.)

imposed by the high risk of artifact complications, however, progress continues in many laboratories. The key to future progress resides in systematic approaches to detect, assess, and minimize perturbations, especially those relating to colloid instability artifact.[1,3] As the field of cell biology has shown, the **absence** of artifact is **not** a requirement for interpreting colloidal phenomena well; one need only achieve a certain **minimum level** of artifact.

To prepare flocs for TEM analysis, the best technique in theory for minimal perturbation is the freeze-fracture technique,[91] whereby one constructs metal replicas of fracture planes taken through vitrified samples, using highly specialized equipment. Examination of such replicas by TEM can provide ultrastructural information on hydrated structures at near-nanometre resolution. Unfortunately, for general descriptive work on heterogeneous flocs, freeze-fracture is usually too costly, too time-consuming, and too limited in its potential for revealing chemical information. Also, the difficulty of processing samples in the field for freeze-fracture is prohibitive, thus imposing an unwanted storage step[1] on its use. Despite these practical objections, freeze-fracture can still be of considerable value. Once one has created a visual reconstruction of a floc by other techniques[1,3] yielding images of high integrity, one can employ freeze-fracture images as a standard for a given floc component whose detailed aspect remains in doubt. Related (and much more practical) techniques of cryotechnology[1,91] are evolving to address specific uses. While they lack the high integrity of preservation known for freeze-fracture, they can be modified for specific sample types to minimize the imposed level of perturbation. There is a potential for nonvitrification cryotechnology to permit analysis of the sequence of events in floc development, even if parts of the sequence are relatively rapid.

Hydrophilic embedding resins[1,19] can help bridge the gap in structural integrity between the best cryotechniques for TEM preparation and techniques for conventional embedding[2,90] in hydrophobic resins (followed by the production of ultrathin sections for examination). The state-of-the-art in the embedding of native, aquatic colloids and colloid systems in hydrophilic (water-miscible) resins is achieved through the use of the melamine resin, Nanoplast FB 101.[92,93] With this embedding technique come several advantages over conventional embedding for the production of ultrathin sections of flocs:

1. It yields an image of a **hydrated** colloid system, including flocs rich in organic materials such as fibrils, even though the embedding matrix becomes dehydrated during the hardening step prior to sectioning.[19,92,93]
2. Shrinkage is minimal, and, as a consequence, so are positional artifacts of differential shrinkage.[92,93]
3. The practical resolution limit is improved.[1,48,92,93]
4. Embedding can be carried out directly in the field with samples as they are taken, thus reducing storage time to zero.[1]
5. Since chemical preservation (fixation) is not a requirement for **noncellular** floc components, the preservation can be carried out with no fixation artifacts for the non-cellular units of structure.[1]

The drawbacks of hydrophilic embedding are few. The technology of selective staining for flocs in a melamine matrix is in its infancy; research in technique modification is improving the situation.[1] In the absence of a fixation step, intracellular structures are **not** perturbed minimally by Nanoplast FB 101. This drawback is readily addressed through split-sample strategies involving a multi-method approach to embedding.[2] Such strategies involve correlating the images of cells embedded in Nanoplast (with and without a prior fixation) and images derived from conventionally-embedded cells whose embedding had been preceded by conventional fixation and dehydration procedures.[2,3]

A new use for Nanoplast preparations has emerged recently, the use of (initially) fluid Nanoplast films for embedding **small** intact flocs directly on a grid used for TEM examination.[1,19,86] With the film technique, the smallest native flocs (submicron flocs) can be added directly to an unpolymerized resin in such a way as to create (after rapid polymerization) a support film whose flocs are embedded within. This support film can be made as thin as is required to achieve a TEM resolution similar to that obtained with 50-nm sections. Flocs which are relatively large with respect to film thickness appear suspended on the grid surface, held in place and coated by electron-transparent matrix polymers. The analysis of the scavenging activity of riverine fibril flocs[94] and of growing mixed aggregates (organic/inorganic associations)[95] is currently being facilitated by the Nanoplast film technique used in the field (see Figure 2).

C. Multi-Method Analysis

The integration of multi-method analyses allows one to surround the problem of artifact, permitting distinctions that no one method can yield unequivocally. If several different chemical/physical/biological techniques are applied to the analysis of floc structure and behavior, and all point to a specific conclusion with regard to a specific phenomenon, then that conclusion is likely to be correct. The blending of multi-method analyses with the minimal perturbation approach has been reviewed recently in the context of aquatic colloid systems.[2,19] A helpful example of a practical strategy for addressing native, aquatic colloid phenomena, based on astute combinations of chemical/limnological analyses in the field, has been developed for describing mineral colloid behavior in organic-rich waters;[96] this strategy, readily coupled with the Nanoplast film technique,[19] is transferable to characterizations of organic flocs and to their activities in surface waters.

By using a judicious selection of electron microscope apparatus[1] and modes for viewing[1] samples, one can greatly enrich the information output of a microscope-based examination. When large numbers of ultrathin sections are required to characterize a given colloid system,[1] it can become profitable to employ high-voltage electron microscopy (HVEM)[97] and/or novel methods of image analysis[1,97] to permit examination of relatively thick sections. A HVEM (the big brother of TEM) can achieve near-nanometre resolution when used with sections which are 10 to 20 times the thickness permitted by a conventional TEM, although not without presenting some technical problems requiring specialized expertise. For flocs with significant mineral components and/or organic components which can be selectively stained with metals,[1] the technology of energy-dispersive spectroscopy[98-100] can help relate chemical information to morphological information. Microchemistry and cytochemistry[1] can provide a similar benefit, as can electron diffraction on a per-colloid basis carried out in the specimen plane of a TEM. Improved techniques for maintaining the integrity of organic crystals during the intensive bombardment necessary for TEM-based electron diffraction are amenable for use with environmental particles in the colloid size range.[101]

D. Coping with Heterogeneity

Aquatic flocs can be highly heterogeneous; some common recognizable constituents are polysaccharide fibrils, humic substances, bacteria, viruses, prokaryote algae (cyanobacteria), small eukaryote algae, clay minerals, iron oxyhydroxides, biogenic silicates, biogenic calcium compounds, bits of organic skeletal materials (e.g., cell wall, algal scale, chitin exoskeleton), tripartite membranes, and transient ultrastructural units released from lysing cells.[1] The literature base used for identifying these morphological entities on the basis of size, shape, internal differentiation, staining characteristics, and X-ray microanalysis has been reviewed recently.[1] This literature base, while generally excellent, is flawed from one major point-of-view: much of it ignores **native**, suspended mixed particles (including flocs), preferring instead to present a specific component in a perturbed state. Minerals are shown in a "cleaned" state, and extracellular biologicals are presented in a dehydrated state (by TEM) without regard to their natural associations. Thus, image interpretation requires a blend of hard science and detective work (the "art" of microscopy), which can cut across many scientific disciplines and diverse technologies.

For assistance in making fine distinctions with TEM images, the reader is directed to:

1. A general atlas of biological ultrastructure[47]
2. A general atlas of mineral ultrastructure[102]
3. A review which relates ultrastructural aspects of the subcellular order within prokaryote cells to the plasticity that such cells exhibit under changing environmental circumstances[103]
4. A review of the strategy and tactics which are available for high-resolution analyses of cell surfaces[104]

The latter review[104] treats perturbing techniques quite extensively, showing how to employ them for detailed characterizations, so as to take artifact properly into account. The use of perturbing "shortcuts" can be justified when a specific material for analysis has been extensively investigated by multimethod approaches in previous researches.[1]

E. Correlative Microscopy

Correlative microscopy is a strategy for utilizing several different kinds of microscopes and accessory techniques (including preparatory techniques) in a multi-method context to analyze a given specimen for different kinds of information. To describe the ultrastructure of a floc, one can profitably combine observations from optical microscopy, scanning electron microscopy (SEM), and TEM, including hybrid (STEM)[105] and derivative TEM instruments (HVEM).[97] Evaluations of correlative microscopy have been published recently with regard to describing aquatic colloid systems[1] and solving particle-speciation problems in aquatic ecosystems.[3] The future of native floc characterization resides in making more extensive use of correlative microscopy in conjunction with the methods of quantitative analytical chemistry.

F. Characterization Data: Minimal vs. Uncontrolled Perturbation

Data derived from particle analysis techniques applied to common aquatic materials can be extremely distorted and misleading when perturbing influences go unrecognized or uncontrolled. Four kinds of apparently innocuous factors which can impinge dramatically on nonliving colloid systems are considered briefly below. These factors are storage time; mode of (harshness of) dehydration; concentration factor; and, for fractionated natural water, the flow rate through a filter-fractionation apparatus.

1. Storage — When pedogenic fulvic acids are stored as a concentrate (>1 g/l) prior to analysis, the individual colloids of ca. 0.002-μm diameter will aggregate irreversibly to form particulates of conventional size (ca. 1-μm diameter and larger).[41] Several investigators have shown that storage of natural water samples for more than one day can produce large changes in the extent of aggregation of suspended colloids.[19]
2. Harsh dehydration — When acid polysaccharide fibrils in lake water are concentrated by bulk freeze-drying, the colloid system changes from one of dispersed fibrils (ca. $5 \times 5 \times 1000$ nm) to one of stable macroscopic fibrous sheets ($>10^5 \times 10^5 \times 10^3$ nm).[2,9]
3. Concentration factor — Within a pH range of 3.5 to 10.0, the structure of fulvic acid aggregates in water is essentially a function of sample concentration,[42] with many different morphological entities extending over much of the colloidal size range.[41,42]
4. Flow rate — The percent of iron colloid (from water rich in organic carbon and iron) retained on the surface of a membrane (in a filtration apparatus) can vary by more than 400% when the flow rate is varied from 1 cm h^{-1} to 20 cm h^{-1}.[48] The mechanisms whereby flow rate contributes to inadvertent aggregation artifact are being investigated intensively for native colloid systems in lake water.[81]

The four factors are all related to the degree of hydration of a floc's polymer matrix. The shrinkage of floc matrix which can occur as a result of **not** controlling the four factors results essentially from a loss of bulk water from the gel structure of the matrix. This, in turn, permits adjacent matrix polymers (usually in the form of fibrils) to adhere to each other, often in an irreversible manner, thus establishing a physical resistance to a full rehydration. Control of these four factors, which perturb the water distribution within a floc, can be a simple matter. The storage time, concentration factor, and flow rate should be minimized to the greatest extent feasible. Dehydration should be avoided, if possible; if not, several methods of dehydration should be compared for perturbing influences detectable by TEM, with the least perturbing method becoming the method of choice. The extra effort devoted to minimal perturbation can certainly be repaid in terms of an improved data collection.[9,19,20,48,55,94]

IV. RECENT EXPERIMENTAL ANALYSES OF UNDEGRADED NATIVE COLLOID SYSTEMS

Making full use of the minimal perturbation approach in conjunction with multi-method analyses is an endeavor still in its infancy. There are, however, at least three natural experimental systems whereby an optimal use can be attempted now. They are (1) organic fibril flocs implicated in riverine contaminant transport (Rhine River, Switzerland, and Germany);[94] (2) iron oxyhydroxyphosphate colloid systems present at the oxic-anoxic interface of organic-rich lake waters (Lake Bret, Vaud, Switzerland);[19,48,106] and (3) colloid systems present in iron- and manganese-rich layers of organic-rich lake sediments (Brady Lake, Ontario, Canada).[19,107,108] Some of the information generated by these systems, for use in interpreting aquatic ecosystem dynamics, was unobtainable by conventional research approaches in use prior to 1988. The main advances came on four fronts: (1) the sizing of individual colloids within a colloid system; (2) descriptions of native associations in heterogeneous aggregates; (3) compositional analyses on a "per-colloid" basis; and (4) the revelation of crystalline regions within materials routinely considered amorphous. A general context for these advances is found in recent reviews.[1,3]

V. IMPLICATIONS OF AN IMPROVED CAPACITY FOR REALISTIC ANALYSES OF FLOCS

A. Research Applications of an Improved Understanding of Floc Structure

The research applications are likely to take two principal directions: (1) learning more about aquatic ecosystems and (2) learning more about the apparatus of separation and fractionation as they apply to sample processing and water decontamination. We need better information about the natural transporters of contaminants in ecosystems so as to have an improved predictive capacity and action plan in the face of a major contamination episode; we need better information about these same transporters to improve our capacity to manipulate them better for optimal water quality in the face of today's multiple uses of (and stresses on) natural water bodies. In this vein, it would be helpful to place more research effort on analyzing the phenomenon of "colloidal pumping",[27] the transfer of sorbed contaminants from the colloid pool to larger aggregates which participate in sedimentation.

B. Literature Values Based on Uncontrolled Colloid Instability Artifact — What Can We Salvage?

A given publication from the pre-1990 literature on suspended organic flocs in surface waters can contain errors of detail, even if the conclusions have subsequently been verified. If a less perturbing technology applied to a given type of water sample puts an investigator in conflict with past data on the same type of sample, it might be wise to consider outright that the past data is incorrect, and then continue onward. Where decisions on sizing were made with a standard cutoff filter[1] (usually 0.45 μm) to distinguish particulate from soluble entities (with no reference to the presence of colloids or monitoring for artifactual aggregation), it is sometimes best to ignore a published work completely. There are reasons why colloidal carriers of contaminants appear as solutes in one laboratory, while appearing as conventional particles in another, thus creating a chaotic situation for those attempting to model contaminant dispersion. The reasons reside in uncontrolled, colloid instability artifact, a situation in urgent need of change.[20,80] Given the increasing cost of environmental research and the increasing capacity of analytical techniques to make fine discriminations, an unaddressed misfractionation of a water sample can lead a research project to financial disaster. The flocs recorded by particle analysis technology should be flocs which actually existed in a water sample; by adopting a firm commitment to this standard, we should advance faster in our understanding of contaminant partitioning into flocs and related occlusion phenomena.

This section would not be complete without specific reference to humic substances and their aggregates, so heavily implicated in contaminant dispersion and impact on biota.[1,31,38] The humic substances (including fulvic acids) used in laboratories to analyze humic/contaminant interactions in water are almost always **degraded** versions of the natural substances employed ostensibly for model system analysis. The time has arrived to move forward from model systems to real systems; there is no reason to believe that a degraded humic substance will behave similarly to a native one, and there is considerable suspicion that some differences will be great.

REFERENCES

1. Leppard, G. G., Evaluation of electron microscope techniques for the description of aquatic colloids, in *Environmental Particles*, Buffle, J. and van Leeuwen, H. P., Eds., IUPAC Environmental Chemistry Series, *Vol. 1*, Lewis Publishers, Chelsea, MI, 1992, chap. 6.
2. Leppard, G. G., Burnison, B. K., and Buffle, J., Transmission electron microscopy of the natural organic matter of surface waters, *Anal. Chim. Acta*, 232, 107, 1990.
3. Leppard, G. G., Size, morphology and composition of particulates in aquatic ecosystems: solving speciation problems by correlative electron microscopy, *Analyst*, 117, 595, 1992.
4. Leppard, G. G., The ultrastructure of lacustrine sedimenting materials in the colloidal size range, *Arch. Hydrobiol.*, 101, 521, 1984.
5. Massalski, A. and Leppard, G. G., Survey of some Canadian lakes for the presence of ultrastructurally discrete particles in the colloidal size range, *J. Fish. Res. Board Can.*, 36, 906, 1979.

6. Beveridge, T. J. and Graham, L. L., Surface layers of bacteria, *Microbiol. Rev.*, 55, 684, 1991.
7. Costerton, J. W., Irvin, R. T., and Cheng, K.-J., The bacterial glycocalyx in nature and disease, *Annu. Rev. Microbiol.*, 35, 299, 1981.
8. Costerton, J. W., Cheng, K.-J., Geesey, G. G., Ladd, T. I., Nickel, J. C., Dasgupta, M., and Marrie, T. J., Bacterial biofilms in nature and disease, *Annu. Rev. Microbiol.*, 41, 435, 1987.
9. Leppard, G. G., The fibrillar matrix component of lacustrine biofilms, *Water Res.*, 20, 697, 1986.
10. Geesey, G. G., Microbial exopolymers: ecological and economic considerations, *ASM News*, 48, 9, 1982.
11. Gregory, J., Fundamentals of flocculation, *CRC Crit. Rev. Environ. Control*, 19, 185, 1989.
12. Vold, M. J. and Vold, R. D., *Colloid Chemistry, The Science of Large Molecules, Small Particles, and Surfaces*, Reinhold, New York, 1964.
13. Stumm, W., and Morgan, J. J., *Aquatic Chemistry - An Introduction Emphasizing Chemical Equilibria in Natural Waters, Second Edition*, Wiley-Interscience, New York, 1981.
14. O'Melia, C. R., Particle-particle interactions, in *Aquatic Surface Chemistry*, Stumm, W. Ed., John Wiley & Sons, New York, 1987, chap. 14.
15. Stumm, W. and O'Melia, C. R., Stoichiometry of coagulation, *J. AWWA*, 60, 514, 1968.
16. Stumm, W., Chemical interaction in partical separation, *Environ. Sci. Technol.*, 11, 1066, 1977.
17. Jullien, R. and Botet, R., *Aggregation and Fractal Aggregates*, World Scientific, Singapore, 1987.
18. Lin, M. Y., Lindsay, H. M., Weitz, D. A., Ball, R. C., Klein, R., and Meakin, P., Universality in colloid aggregation, *Nature*, 339, 360, 1989.
19. Perret, D., Leppard, G. G., Muller, M., Belzile, N., De Vitre, R., and Buffle, J., Electron microscopy of aquatic colloids: non-perturbing preparation of specimens in the field, *Water Res.*, 25, 1333, 1991.
20. Buffle, J., Leppard, G. G., De Vitre, R. R., and Perret, D., Submicron sized aquatic compounds: from artefacts to ecologically meaningful data, *Proc. Am. Chem. Soc.*, 30(1), 337, 1990.
21. O'Melia, C. R., Kinetics of colloid chemical processes in aquatic systems, in *Aquatic Chemical Kinetics — Reaction Rates of Processes in Natural Waters*, Stumm, W., Ed., John Wiley & Sons, New York, 1990, chap. 16.
22. Weilenmann, U., O'Melia, C. R., and Stumm, W., Particle transport in lakes: models and measurements, *Limnol. Oceanogr.*, 34, 1, 1989.
23. Sigleo, A. C. and Means, J. C., Organic and inorganic components in estuarine colloids: implications for sorption and transport of pollutants, *Rev. Environ. Contam. Toxicol.*, 112, 123, 1990.
24. Decho, A. W., Microbial exopolymer secretions in ocean environments: their role(s) in food webs and marine processes, *Oceanogr. Mar. Biol. Annu. Rev.*, 28, 73, 1990.
25. Sigg, L., Surface chemical aspects of the distribution and fate of metal ions in lakes, in *Aquatic Surface Chemistry*, Stumm, W., Ed., John Wiley & Sons, New York, 1987, chap. 12.
26. Int. Assoc. Great Lakes Res., Eds., *Ecosystems: Contaminants and Surface Films*, Special Issue of *J. Great Lakes Res.*, 8(2), 1982.

27. Honeyman, B. D. and Santschi, P. H., The role of particles and colloids in the transport of radionuclides and trace metals, in *Environmental Particles*, Buffle, J. and van Leeuwen, H. P., Eds., IUPAC Environmental Chemistry Series, *Vol. 1*, Lewis Publishers, Chelsea, MI, 1992, chap. 10.
28. Morel, F. M. M. and Gschwend, P. M., The role of colloids in the partitioning of solutes in natural waters, in *Aquatic Surface Chemistry*, Stumm, W., Ed., John Wiley & Sons, New York, 1987, chap. 15.
29. Ali, W., O'Melia, C. R., and Edzwald, J. K., Colloidal stability of particles in lakes: measurement and significance, *Water Sci. Tech.*, 17, 701, 1985.
30. O'Melia, C. R., The influence of coagulation and sedimentation on the fate of particles, associated pollutants, and nutrients in lakes, in *Chemical Processes in Lakes*, Stumm, W., Ed., Wiley-Interscience, New York, 1985, 207.
31. Suffet, I. H. and MacCarthy, P., Eds., *Aquatic Humic Substances — Influence on Fate and Treatment of Pollutants*, American Chemical Society, Washington, D.C., 1989.
32. Brown, M. J. and Lester, J. N., Metal removal in activated sludge: the role of bacterial extracellular polymers, *Water Res.*, 13, 817, 1979.
33. Van Loosdrecht, M. C. M., Lycklema, J., Norde, W., and Zehnder, A. J. B., Influences of interfaces on microbial activity, *Microbiol. Rev.*, 54, 75, 1990.
34. Hunter, K. A. and Liss, P. S., Organic sea surface films, in *Marine Organic Chemistry — Evolution, Composition, Interactions and Chemistry of Organic Matter in Seawater*, Duursma, E. K. and Dawson, R., Eds., Elsevier, Amsterdam, 1981, chap. 9.
35. Newton, P. P., and Liss, P. S., Surface charge characteristics of oceanic suspended particles, *Deep-Sea Res.*, 36, 759, 1989.
36. Leppard, G. G., Organic coatings on suspended particles in lake water, *Arch. Hydrobiol.*, 102, 265, 1984.
37. Gibbs, R. J., Effect of natural organic coatings on the coagulation of particles, *Environ. Sci. Technol.*, 17, 237, 1983.
38. Buffle, J., *Complexation Reactions in Aquatic Systems: An Analytical Approach*, Ellis Horwood, Chichester, England, 1988.
39. Leppard, G. G., Massalski, A., and Lean, D. R. S., Electron-opaque microscopic fibrils in lakes: their demonstration, their biological derivation and their potential significance in the redistribution of cations, *Protoplasma*, 92, 289, 1977.
40. Caceci, M. S. and Billon, A., Evidence for large organic scatterers (50-200 nm diameter) in humic acid samples, *Org. Geochem.*, 15, 335, 1990.
41. Leppard, G. G., Buffle, J., and Baudat, R., A description of the aggregation properties of aquatic pedogenic fulvic acids — combining physico-chemical data and microscopical observations, *Water Res.*, 20, 185, 1986.
42. Stevenson, I. L. and Schnitzer, M., Transmission electron microscopy of extracted fulvic and humic acids, *Soil Sci.*, 133, 179, 1982.
43. Gschwend, P. M. and Wu, S.-C., On the constancy of sediment-water partition coefficients of hydrophobic organic pollutants, *Environ. Sci. Technol.*, 19, 90, 1985.
44. Honeyman, B. D. and Santschi, P. H., Metals in aquatic systems, *Environ. Sci. Technol.*, 22, 862, 1988.
45. Beveridge, T. J., Role of cellular design in bacterial metal accumulation and mineralization, *Annu. Rev. Microbiol.*, 43, 147, 1989.
46. Beveridge, T. J., Ultrastructure, chemistry, and function of the bacterial wall, *Int. Rev. Cytol.*, 72, 229, 1981.

47. Lima-de-Faria, A., Ed., *Handbook of Molecular Cytology*, North-Holland, Amsterdam, 1969.
48. Leppard, G. G., De Vitre, R. R., Perret, D., and Buffle, J., Colloidal iron oxyhydroxy-phosphate: the sizing and morphology of an amorphous species in relation to partitioning phenomena, *Sci. Tot. Environ.*, 87/88, 345, 1989.
49. Strycek, T., Acreman, J., Kerry, A., Leppard, G. G., Nermut, M. V., and Kushner, D. J., Extracellular fibril production by freshwater algae and cyanobacteria, *Microbial Ecol.*, 23, 53, 1992.
50. Ramamoorthy, S. and Leppard, G. G., Fibrillar pectin and contact cation exchange at the root surface, *J. Theor. Biol.*, 66, 527, 1977.
51. Marshall, K. C., Stout, R., and Mitchell, R., Mechanism of the initial events in the sorption of marine bacteria to surfaces, *J. Gen. Microbiol.*, 68, 337, 1971.
52. Wimpenny, J. W. T., Coombs, J. P., and Lovitt, R. W., Growth and interactions of microorganisms in spatially heterogeneous ecosystems, in *Current Perspectives in Microbial Ecology*, Klug, M. J. and Reddy, C. A., Eds., American Society for Microbiology, Washington, D.C., 1984, 291.
53. Paul, J. H. and Jeffrey, W. H., Evidence for separate adhesion mechanisms for hydrophilic and hydrophobic surfaces in *Vibrio proteolytica*, *Appl. Environ. Microbiol.*, 50, 431, 1985.
54. Shea, C., Nunley, J. W., Williamson, J. C., and Smith-Somerville, H. E., Comparison of the adhesion properties of *Deleya marina* and the exopolysaccharide-defective mutant strain DMR, *Appl. Environ, Microbiol.*, 57, 3107, 1991.
55. Burnison, B. K. and Leppard, G. G., Isolation of colloidal fibrils from lake water by physical separation techniques, *Can. J. Fish. Aquat. Sci.*, 40, 373, 1983.
56. Friedman, B. A. and Dugan, P. R., Concentration and accumulation of metallic ions by the bacterium *Zoogloea*, *Dev. Ind. Microbiol.*, 9, 381, 1968.
57. Friedman, B. A., Dugan, P. R., Pfister, R. M., and Remsen, C. C., Fine structure and composition of the zoogloeal matrix surrounding *Zoogloea ramigera*, *J. Bacteriol.*, 96, 2144, 1968.
58. Friedman, B. A., Dugan, P. R., Pfister, R. M., and Remsen, C. C., Structure of exocellular polymers and their relationship to bacterial flocculation, *J. Bacteriol.*, 98, 1328, 1969.
59. Leshniowsky, W. O., Dugan, P. R., Pfister, R. M., Frea, J. I., and Randles, C. I., Adsorption of chlorinated hydrocarbon pesticides by microbial floc and lake sediment and its ecological implications, in *Proc. 13th Conf. Great Lakes Res., Part 2*, Int. Assoc. Great Lakes Res., Eds., Braun-Brumfield, Ann Arbor, MI, 1970, 611.
60. Leshniowsky, W. O., Dugan, P. R., Pfister, R. M., Frea, J. I., and Randles, C. I., Aldrin: removal from lake water by flocculent bacteria, *Science*, 169, 993, 1970.
61. Means, J. C. and Wijayaratne, R., Role of natural colloids in the transport of hydrophobic pollutants, *Science*, 215, 968, 1982.
62. Voice, T. C., Rice, C. P., and Weber, W. J., Effect of solids concentration on the sorptive partitioning of hydrophobic pollutants in aquatic systems, *Environ. Sci. Technol.*, 17, 513, 1983.
63. Wijayaratne, R. D. and Means, J. C., Affinity of hydrophobic pollutants for natural estuarine colloids in aquatic environments, *Environ. Sci. Technol.*, 18, 121, 1984.
64. Freeman, D. H. and Cheung, L. S., A gel partition model for organic desorption from a pond sediment, *Science*, 214, 790, 1981.

65. Means, J. C. and Wijayaratne, R. D., Sorption of benzidine, toluidine, and azobenzene on colloidal organic matter, in *Aquatic Humic Substances — Influence on Fate and Treatment of Pollutants*, Suffet, I. H. and MacCarthy, P., Eds., American Chemical Society, Washington, D.C., 1989, chap. 14.
66. Pfister, R. M., Dugan, P. R., and Frea, J. I., Microparticulates: isolation from water and identification of associated chlorinated pesticides, *Science*, 166, 878, 1969.
67. Yin, C. and Hassett, J. P., Fugacity and phase distribution of Mirex in Oswego River and Lake Ontario waters, *Chemosphere*, 19, 1289, 1989.
68. Wijayaratne, R. D. and Means, J. C., Sorption of polycyclic aromatic hydrocarbons by natural estuarine colloids, *Mar. Environ. Res.*, 11, 77, 1984.
69. Brownawell, B. J. and Farrington, J. W., Biogeochemistry of PCBs in interstitial waters of a coastal marine sediment, *Geochim. Cosmochim. Acta*, 50, 157, 1986.
70. Brannon, J. M., Myers, T. E., Gunnison, D., and Price, C. B., Nonconstant polychlorinated biphenyl partitioning in New Bedford Harbor sediment during sequential batch leaching, *Environ. Sci. Technol.*, 25, 1082, 1991.
71. Albergoni, V. and Piccinni, E., Biological response to trace metals and their biochemical effects, in *Trace Element Speciation in Surface Waters and its Ecological Implications*, Leppard, G. G., Ed., Plenum Press, New York, 1983, 159.
72. Cowen, J. P. and Silver, M. W., The association of iron and manganese with bacteria on marine macroparticulate material, *Science*, 224, 1340, 1984.
73. Whitfield, C., Bacterial extracellular polysaccharides, *Can. J. Microbiol.*, 34, 415, 1988.
74. Hunter, K. A., Microelectrophoretic properties of natural surface-active organic matter in coastal seawater, *Limnol. Oceanogr.*, 25, 807, 1980.
75. Hunter, K. A., The adsorptive properties of sinking particles in the deep ocean, *Deep-Sea Res.*, 30, 669, 1983.
76. Chase, R. R. P., Settling behavior of natural aquatic particulates, *Limnol. Oceanogr.*, 24, 417, 1979.
77. Martin, J. M. and Moreira-Turcq, P., A new methodology for characterizing organic coatings of aquatic particles, *Mar. Pollut. Bull.*, 22, 287, 1991
78. Sonnenfeld, E. M., Beveridge, T. J., and Doyle, R.J., Discontinuity of charge on cell wall poles of *Bacillus subtilis*, *Can. J. Microbiol.*, 31, 875, 1985.
79. Mayers, I. T. and Beveridge, T. J., The sorption of metals to *Bacillus subtilis* walls from dilute solutions and simulated Hamilton Harbour (Lake Ontario) water, *Can. J. Microbiol.*, 35, 764, 1989.
80. Buffle, J., Perret, D., and Newman, M., The use of filtration and ultrafiltration for size fractionation of aquatic particles, colloids and macromolecules, in *Environmental Particles*, Buffle, J. and van Leeuwen, H. P. Eds., IUPAC Environmental Chemistry Series, *Vol. 1*, Lewis Publishers, Chelsea, MI, 1992, chap. 5.
81. Perret, D., De Vitre, R. R., Leppard, G. G., and Buffle, J., Characterizing autochthonous iron particles and colloids — the need for better particle analysis methods, in *Large Lakes: Ecological Structure and Function*, Tilzer, M. M. and Serruya, C., Eds., Springer-Verlag, Berlin, 1990, 224.
82. Hirtzel, C. S., and Rajagopalan, R., *Colloidal Phenomena: Advanced Topics*, Noyes Publishers, Park Ridge, N.J, 1985.
83. Prost, J. and Rondelez, F., Structures in colloidal physical chemistry, *Nature*, Suppl. 350, 11, 1991.

84. Wiggins, P. M., Role of water in some biological processes, *Microbiol. Rev.*, 54, 432, 1990.
85. Saenger, W., Structure and dynamics of water surrounding biomolecules, *Ann. Rev. Biophys. Biophys. Chem.*, 16, 93, 1987.
86. Leppard, G. G., Belzile, N., Perret, D., Filella, M., and Buffle, J., unpublished results, 1991.
87. Leppard, G. G., Burnison, B. K., and Carey, J. H., unpublished results, 1991.
88. Costerton, J. W., Mechanisms of microbial adhesion to surfaces — direct ultrastructural examination of adherent bacterial populations in natural and pathogenic ecosystems, in *Current Perspectives in Microbial Ecology*, Klug, M. J. and Reddy, C. A., Eds., American Society Microbiology, Washington, D.C., 1984, 115.
89. Chapman, S. K., Artifacts of transmission and scanning-transmission electron microscope operation, in *Artifacts in Biological Electron Microscopy*, Crang, R. F. E. and Klomparens, K. L., Eds., Plenum Press, New York, 1988, chap. 7.
90. Hayat, M. A., *Basic Techniques for Transmission Electron Microscopy*, Academic Press, Orlando, 1986.
91. Robards, A. W. and Sleytr, U. B., *Low Temperature Methods in Biological Electron Microscopy*, Elsevier, Amsterdam, 1985.
92. Frosch, D. and Westphal, C., Melamine resins and their application in electron microscopy, *Electron Microsc. Rev.*, 2, 231, 1989.
93. Bachhuber, K. and Frosch, D., Melamine resins, a new class of water-soluble embedding media for electron microscopy, *J. Microsc.*, 130, 1, 1983.
94. Filella, M., Buffle, J., and Leppard, G. G., Characterization of submicron colloids in freshwaters: evidence for their bridging by organic structures, *Water Sci. Technol.*, in press.
95. Pizarro, J., Belzile, N., Filella, M., Leppard, G. G., Perret, D., and Buffle, J., Factors affecting coagulation/sedimentation of lake-borne submicron iron oxyhydroxide particles, submited, 1992.
96. Buffle, J., De Vitre, R. R., Perret, D., and Leppard, G. G., Combining field measurements for speciation in non perturbable water samples, in *Metal Speciation: Theory, Analysis and Application*, Kramer, J. R. and Allen, H. E. Eds., Lewis Publishers, Chelsea, MI, 1988, 99.
97. Turner, J. N., National resources — introduction (special feature of the Electron Microscopy Society of America on high-voltage electron microscopy and HVEM facilities — a series of brief articles by many authors, 104) *EMSA Bull.*, 20 (1), 104, 1990.
98. Warley, A., Standards for the application of X-ray microanalysis to biological specimens, *J. Microsc.*, 157, 135, 1990.
99. Hall, T. A., Quantitative electron probe X-ray microanalysis in biology, *Scanning Microsc.*, 3, 461, 1989.
100. Chandler, J. A., *X-ray Microanalysis in the Electron Microscope*, North-Holland, Amsterdam, 1977.
101. Fryer, J. R. and Dorset, D. L., Eds., *Electron Crystallography of Organic Molecules*, Kluwer Academic Press, Dordrecht, The Netherlands, 1991.
102. Beutelspacher, H. and van der Marel, H. W., *Atlas of Electron Microscopy of Clay Minerals and their Admixtures: A Picture Atlas*, Elsevier, Amsterdam, 1968.
103. Costerton, J. W., Structure and plasticity at various organization levels in the bacterial cell, *Can. J. Microbiol.*, 34, 513, 1988.

104. Nermut, M. V., Strategy and tactics in electron microscopy of cell surfaces, *Electron Microsc. Rev.*, 2, 171, 1989.
105. Brown, L. M., Scanning transmission electron microscopy: microanalysis for the microelectronic age, *J. Phys.*, F11, 1, 1981.
106. Leppard, G. G., Buffle, J., De Vitre, R. R., and Perret, D., The ultrastructure and physical characteristics of a distinctive colloidal iron particulate isolated from a small eutrophic lake, *Arch. Hydrobiol.*, 113, 405, 1988.
107. De Vitre, R., Belzile, N., Leppard, G. G., and Tessier, A., Diagenetic manganese and iron oxyhydroxide particles collected from a Canadian lake: morphology and chemical composition, in *Heavy Metals in the Environment, Vol. 1*, Vernet, J. P. Ed., CEP Consultants, Edinburgh, 1989, 217.
108. Belzile, N., De Vitre, R. R., and Tessier, A., *In situ* collection of diagenetic iron and manganese oxyhydroxides from natural sediments, *Nature*, 340, 376, 1989.

CHAPTER 8

Suspended Particle Research: Current Approaches and Future Needs

Terrance J. Beveridge, Salem S. Rao, Sherland A. Daniels, and Colin M. Taylor

TABLE OF CONTENTS

0-87371-678-7/93/$0.00+$.50

I. INTRODUCTION

There are a variety of particles within the aquatic milieu. The smallest component is a "primary particle", which can be organic (e.g., bacteria) or inorganic (e.g., clay) and small enough to be colloidal (e.g., <1.0 μm in the least dimension of the particle). Under natural conditions, these primary particles usually adhere together to form larger particles or "flocs" constituting composites of several different organic/inorganic primary particles, which makes them extremely complicated heteromaterials. They exhibit variable shapes and topography, a mixture of available chemical sites in varying proportions, and varying degradation rates. Furthermore, the weak attractive forces which hold together the particles of a floc can be readily broken, thus ensuring that each floc is a dynamic aggregate, continuously breaking down or building up, depending on environmental parameters such as shear stress.[1] Flocs present complex properties which are poorly understood, and which will affect particle residence time in the aquatic environment, as well as the ability to interact with and bind contaminants within a freshwater ecosystem.[2,3] More information is required about the formation of natural flocs, their dynamical properties, their chemical constitution, their degradation rates, and their ability to "fix" environmental contaminants. In order to recognize their remobilization potential, we must also attempt to understand the tenaciousness of contaminant fixation.

II. BACTERIA AND SUSPENDED PARTICULATES

In freshwater ecosystems, it is well known that suspended particles play an important role in contaminant binding and transport.[4] There are several factors which influence this: particle nature and size distribution, mineralogy, surface area, and chemical coating.[4-9] Fletcher and Floodgate[10] and Rao et al.[11] suggested that adhesion of bacteria to surfaces may enhance the production of large aquatic aggregates, which would, in turn, affect the harboring of contaminants. In fact, data continue to accumulate which suggest that, in nature, clean mineral faces simply do not exist. Interfacial forces attract and concentrate most solutes, and microbes adhere to mineral faces to feed on the collected organics. These phenomena represent early stages in the development of the so-called "biofilm".

Before attempting to describe the complexity involved in the formation of microbial biofilms and the eventual production of large contaminant-laden aggregate matrices, it is desirable to first outline how bacterial surfaces interact with abiotic particles. Most of the available information regarding this work has had a strong input from chemists who have helped us understand abiotic surface properties and floc formation[12-15] as well as from contributions of engineers, geomorphologists, and geographers.[16-21]

In marine environments, where most studies were primarily conducted, there is a strong enhancement of electrochemical flocculation which is caused by the electrical double-layer effect on surfaces being augmented by a high salt content in the water. Without such a strong enhancement in freshwater, the extent of electrochemically-based flocculation was considered to be less than what recent studies have indicated.[22,23,11] Bactera[3] and their cell-wall exudates[26] exist in both freshwater and marine environments,[24] and have been found to play a role in the process of heavy-metal contaminant binding.

The rationale behind this bacterial line of investigation is clearly outlined.[2] Bacteria have the highest surface area to volume ratio of all cells, and, consequently, they have extensive interactions with their aqueous surroundings as well as with their neighboring cells. Most of the prokaryotic surfaces studied so far (including cell walls and their external polymers) possess an overall electronegative charge.[2] This suggests that they have a high capacity to sorb large quantities of metals onto their surfaces; so much so that, in time, small-grained minerals are formed and their metal constituents are effectively immobilized.[3] For cell-to-cell interactions, it is very possible that multivalent metallo-ions (e.g., Mg^{2+}, Ca^{2+}, Fe^{3+}, etc.) could act as convenient agents to cement cells together by salt-bridging. Monovalent ions (e.g., Na^{+}, K^{+}, etc.), on the other hand, could neutralize polar sites, increase surface hydrophobicity, and anneal cells together by hydrophobic interactions. The same holds true for their interaction with small-grained solids such as clays.[26] It is known that clay-bacterial composites are common to most aquatic environments.[3] It has also been suggested that bacterial surfaces are not homogeneous with regard to charge distribution. Instead, they are more like patchwork quilts composed of an amalgam of different charge intensities.[27]

It is, therefore, reasonable to expect that the chemical forces which attach neighboring bacterial cells together and which cause interactions with clays and other minerals in aquatic systems would be the result of electrostatic, salt-bridging, and hydrophobic interactions of variable intensity over the entire cell surface. The increase in the size of flocs is time dependent. It is a function of such diverse factors as ionic concentration and species, temperature, pH, turbulence, and shear force. It is highly probable that small-grained solids would also become entrapped within a consolidating floc matrix. Since flocs seem to be a natural phenomenon of both marine as well as freshwater ecosystems, and since microbial metabolic transformations can affect water quality, these fundamental surface traits responsible for floc formation deserve greater consideration.

III. COLLOIDS

Although often overlooked, "colloids" (operationally defined as small particles of variable composition) are also very important in aquatic environments where they can act as contaminant carriers. Colloidal particles are extremely plentiful and small, being no more than 1.0 μm in their least dimension, and easily remain in fluid suspension. They contribute to the overall aggregate structure as adhesives and contain adsorption sites. These particles must be potential concentrators of aquatic contaminants.

Common organic substances contributing to colloids in aquatic systems are (1) humic acid polymers, which are a poorly defined group of organo-heteropolymers possessing both negatively-charged and hydrophobic sites, and (2) acidic polysaccharides. Both classes of polymers are interactive with metal ions to produce organo-metallic complexes. The importance of the humic polymers was first postulated by Karickhoff et al.[28] and Karickhoff,[29] who worked with bottom sediments. Little is known about the role that these colloids play in toxic contaminant transport, but indirect evidence suggests their importance.[28] This knowledge gap is largely attributed to the difficulties in both separating and chemically characterizing them. Attempts, however, have been made by Burnison and Leppard[31] and Leppard[30] to isolate and characterize acidic polysaccharides using electron microscopy and microchemical techniques. They found that this material (<0.45 μm at least diameter; collected by the Hollow Fibre Ultra Filtration Technique), especially in the larger fractions, consisted of fibrillar colloids with the fibrils resembling components of bacterial capsules. These tiny, 0.005 μm in diameter, particles contributed to the overall production of large aquatic suspended particulates,[32,33] and, hence, must be considered in studies involving interactions between contaminants and suspended particles.

Submicroscopic analysis, employing correlative electron microscopy, energy dispersive spectroscopy, and microchemical techniques, has been used to characterize both organic and inorganic colloid-particle complexes. Not only have these studies supported the notion that colloids are involved in contaminant binding and contaminant transport,[34] but they have also overcome the microscopical and simple-handling artifact problems previously encountered.[30,35]

It is apparent from the literature that two independent approaches (physico-chemical and biological) have been adopted to understand the complex behavior of environmental contaminants in aquatic ecosystems. The two approaches rarely congeal during the studies of aquatic contaminant behavior, and it is time for a more interdisciplinary attack.

IV. PHYSICAL AND CHEMICAL STUDIES

The rationale normally taken by physical modelers is to develop contaminant transport models based on the knowledge that a large portion of the contaminants is predominantly associated with the stable suspended particles and the bottom sediments. This was supported by observations made by chemists, where the concentration of specific contaminants was measured in both suspended particles and bottom sediments of various rivers. These studies did not consider the potential of the delicate, but important, bacterial interplay with primary particles (or those of colloidal materials) to help form aggregates (suspended particulates), or their role in contaminant binding and subsequent transport.

Indeed, physical modelers believed that cohesive sediment particles (<60 µm-sized particles) were the primary carriers of aquatic contaminants. A review of their work relating to sediment-associated contaminant transport was reviewed by Krishnappan and Ongley,[36] and revealed that considerable progress has been made in the development of fine-grained sediment computer models. The list of computer programs includes: HEC-6 (1977), IALLUVIAL 11 (1982), MOBED (1981), and FLUVIAL 11 (1982), which are extensively used to predict sediment movements. These models are not capable, however, of coping with the transport of fine-grained cohesive sediments, which is very important in the actual prediction of events within rivers. Krishnappan and Ongley[36] described the importance of these fine sediments in the transportation of chemical contaminants. They recognized that the mechanisms controlling the formation and transportation of suspended aggregates are far more complex than originally expected. Furthermore, the issue of the transport of organic contaminants becomes particularly significant in view of the importance of toxicants/contaminants associated with biological entities (especially bacteria) that are themselves associated with cohesive particles in the water column.[11,37] Indeed, some organic contaminants are not only sorbed to these particles, but they are also chemically modified by the metabolic processes of the microorganisms.

V. BACTERIAL BIOSORPTION OF ORGANIC CONTAMINANTS

Steen and Karickhoff[38] conducted a study of biosorption by sediment/soil-derived microbial populations of pyrene and phenanthrene, in which they found that the measured biosorption coefficients of these two environmentally relevant contaminants appeared to be similar for microbial populations, regardless of their sources. Furthermore, Baughman and Paris[39] cited several sources to show that the sorption of organic pollutants occurs, at least to the same extent, in dead cells as in living cells. Certain toxicants can be detoxified and can even serve as food sources for microbes. In addition, Grimes and Morrison[40] investigated the prevalence of uptake of the chlorinated hydrocarbon insecticides (CHI), such as α- and γ-9 Chlordane, dieldrin, heptachlor epoxide, and lindane, from aquatic environments by chemoorganotrophic bacteria. They found the uptake of CHI to be rapid, with a near maximum quantity being sorbed within 15 min. These CHI were not easily removed (desorbed) from the cells, and desorption was directly proportional to contaminant/water solubility. They even suggested that bioconcentration of CHI by aquatic bacteria might serve as a means of introducing these toxic compounds into the aquatic food chain. They felt that the process of microbial uptake and bioconcentration might be adapted as a treatment procedure for the intentional removal of the contaminant residue from aquatic environments, wastewater, and water supplies.

In recent studies, Rao et al.[22,41] found that bacteria-rich suspended aggregates have an impact on aquatic environments by controlling the mode of contaminant dispersion. Daniels and Rao[42] used living and dead native, aquatic bacteria to demonstrate an initial rapid adsorption of the contaminant phenanthrene. The initial bacterial adsorption was approximately 85% in experimental units containing living cells or dead bacteria, each at a level of approximately 10^4 cells per milliliter. No further adsorption was demonstrated by the two groups of native bacteria during the 72-h contaminant contact period. However, in experimental units where bacterial populations had been augmented to a level of approximately 10^7 to 10^8 cells per milliliter by organic enrichment, contaminant adsorption was approximately 94%.

These findings suggest increased contaminant sorption by suspended bacterial particulates in circumstances where bacterial concentrations exceed normal levels. These data also agree with the contention of Baughman and Paris[39], and may, in turn, explain why increased levels of bound organic contaminants occur in areas where increased sewage discharges are encountered. Moreover, other published data[43] suggest that the development of suspended bacterial aggregates and their affinity for contaminants have important implications in aquatic environments. Aggregates can also represent a sizeable volume of the suspended material.[16]

Microbiologists, considering the physicochemical properties of bacterial surfaces and bacterial extracelluar exudates associated with metal-binding phenomena, are convinced that bacteria are important in binding and transporting toxic heavy-metals and, presumably, other environmental contaminants. A fuller understanding of the behavior of fine-grained sediment-bacterial complexes will, therefore, advance our knowledge of aquatic contaminant transport models. Hence, the roles played by colloids, bacteria, other abiotic particles and suspended particle-bacterial complexes are all integral elements in the comprehensive study of contaminants in aquatic ecosystems.

We have provided information herein to support the argument that an integrated multidisciplinary approach is needed to understand the complex behavior of environmental contaminants. By combining resources and expertise, therefore, the "physicochemical" and "biological" investigators can perhaps produce a degree of synergism to jointly achieve this mutual objective.

REFERENCES

1. McCoy, W. F., Bryers, J. D., Robbins, J., and Costerton, J. W., Observation on fouling biofilm formation, *Can. J. Microbiol.,* 27, 910, 1981.
2. Beveridge, T. J., The bacterial surface: general considerations towards design and function, *Can. J. Microbiol.,* 34, 363, 1988.
3. Beveridge, T. J., Role of cellular design in bacterial metal accumulation and mineralization, *Annu. Rev. Microbiol.,* 43, 147, 1989.
4. Blachford, D. P. and Day, T. J., Sediment Water Quality Assessments: Opportunities for Integrating Water Quality and Water Resources Branches' Activities, Sediment Survey Sec., IWD-HQ-WRB-SS-88-2, Environment Canada, 1988.

5. Ackermann, F., Bergmann, H., and Schleichert, U., Monitoring of Heavy Metals in Coastal and Estuarine Sediments — A Question of Grain-Size: <20 μm versus <60 μm, *Environ. Tech. Lett.,* 4, 317, 1983.
6. Ongley, E. D., Bynoe, M. C., and Percival, J. B., Physical and geochemical characteristics of suspended solids, Wilton Creek, Ontario, *Can. J. Earth Sci.,* 18, 1365, 1981.
7. Hargrave, B. T. and Kranck, K., Adsorption and transport of pollutants on suspended particles, in Proc. Symp. Non-Biological Transport and Transformation of Pollutants on Land and Water. Processes and Critical Data Required for Predictive Description, National Bureau of Standards, Gaithersburg, MD, 1976.
8. Pfister, R. M., Dugan, P. R., and Frea, J. I., Microparticulates: isolation from water and identification of associated chlorinated pesticides, *Science,* 166, 878, 1969.
9. Lotse, E. G., Graetz, D. A., Chesters, G., Lee, G. B., and Newland, L. W., Lindane adsorption by lake sediments, *Environ. Sci. Tech.,* 2, 353, 1968.
10. Fletcher, M. and Floodgate, G. D., An electron microscopic demonstration of an acidic polysaccharide involved in adhesion of a marine bacterium to solid surfaces, *J. Gen. Microbiol.,* 74, 325, 1973.
11. Rao, S. S., Droppo, I. G., Taylor, C. M., and Burnison, B. K., Fresh Water Bacterial Aggregate Development: Effect of Dissolved Organic Matter, National Water Research Institute Contribution No. 91–75, Canada Centre for Inland Waters, Burlington, Ontario, Canada, 1991.
12. Luckham, P. F. and Vincent, F., The controlled flocculation of particulate dispersions using small particles of opposite charge. II. Investigation of floc structure using a freeze-fracture technique, *Colloids and Surfaces,* 6, 83, 1983.
13. Hunt, J. R., Predictions of oceanic particle size distributions from coagulation and sedimentation mechanisms, in *Particulates in Water: Characterization, Fate, Effects, and Removal. Advances in Chemistry Series 189,* Kavanaugh, M. C., and Leckie, J. O., Eds., American Chemical Society, Washington, D.C., 1980.
14. Sholkovitz, E. R., Flocculation of dissolved organic and inorganic matter during the mixing of river water and seawater, *Geochim. Cosmochim. Acta,* 40, 831, 1976.
15. Allan, R. J., The Role of Particulate Matter in the Fate of Contaminants in Aquatic Ecosystems. Inland Waters Directorate, Scientific Series No. 86–142, National Water Research Institue, Canada Centre for Inland Waters, Burlington, Ontario, Canada, 1986.
16. Droppo, I. G. and Ongley, E. D., Flocculation of suspended solids in southern Ontario rivers, *Sediment and Environment,* Hadley, R. J. and Ongley, E. D., Eds., Int. Assoc. Hydrologic Sciences, 3rd Scientific Assembly, Baltimore, MD, May 10–19, 1989, IAHS Pub. No. 1984.
17. Partheniades, E., The present state of knowledge and needs for future research on cohesive sediment dynamics, Proc. 3rd Int. Symp. River Sedimentation, University of Mississippi, University, 1986. 3.
18. Kranck, K., Dynamics and distribution of suspended particulate matter in the St. Lawrence Estuary, *Naturaliste Can.* 96, 163, 1979.
19. Kranck, K., Particulate matter grain-size characteristics and flocculation in a partially mixed estuary, *Sedimentology,* 28, 107, 1981.

20. Kranck, K., The role of flocculation in the filtering of particulate matter in estuaries, in *The Estuary as a Filter,* Kennedy, V. S., Ed., Academic Press, New York, 1984, 159.
21. Krone, R. B., Aggregation of suspended particles in estuaries, in *Estuarine Transport Processes,* Kjerfve, B., Ed., University of South Carolina Press, Columbia, 1978, 177.
22. Rao, S. S., Weng, J. H., and Krishnappan, B. G., Particle Associated Contaminant Transport in the Yamaska River, National Water Research Institute Contribution No. 88–104, Canada Centre for Inland Waters, Burlington, Ontario, Canada, 1988.
23. Rao, S. S. and Kwan, K. K., Method for Measuring Toxicity of Suspended Particulates in Water, National Water Research Institute Contribution No. 89–102, Canada Centre for Inland Waters, Burlington, Ontario, Canada, 1989.
24. Paerl, H. W., Bacterial uptake of dissolved organic matter in relation to detrital aggregation in marine and freshwater systems, *Limnol. Oceanogr.,* 19, 966, 1974.
25. Geesey, G. G., Microbial exopolymers: ecological and economic considerations, *ASM News,* 49, 9, 1982.
26. Walker, S. G., Flemming, C. A., Ferris, F. G., Beveridege, T. J., and Bailey, G. W., Physicochemical interactions of *Escherichia coli.* Envelopes and *Bacillus subtilis* cell walls with two clays and the ability of the composites to immobilize heavy metals from solution, *Appl. Environ. Microbiol.,* 55, 2976, 1989.
27. Sonnenfeld, E. M., Beveridge, T. J., and Doyle, R. J., Discontinuity of charge on cell wall poles of *Bacillus subtilis, Can. J. Microbiol.,* 31, 875, 1985.
28. Karickhoff, S. W., Brown, D. S., and Scott, T. A., Sorption of hydrophobic pollutants on natural sediments, *Water Res.,* 13, 241, 1979.
29. Karickhoff, S. W., Semi-empirical estimation of sorption of hydrophobic pollutants on natural sediments and soils, *Chemosphere,* 10, 833, 1981.
30. Leppard, G. G., Evaluation of electron microscope techniques for the description of aquatic colloids, in *Environmental Particles,* Buffle, J. and van Leeuwen, H. P., Eds., IUPAC Environmental Chemistry Series, Vol. 1, Lewis Publishers, Chelsea, MI, 1992, 6.
31. Burnison, B. K. and Leppard, G. G., Isolation of colloidal fibrils from lake water by physical separation techniques, *Can. J. Fish. Aquat. Sci.,* 40, 373, 1983.
32. Leppard, G. G., Organic coatings on suspended particles in lake water, *Arch. Hydrobiol.,* 102, 265, 1984.
33. Leppard, G. G., The ultrastructure of lacustrine sedimenting materials in the colloidal size range, *Arch. Hydrobiol.,* 101, 521, 1984.
34. Leppard, G. G., Relationships Between Fibrils, Colloids, Chemical Speciation and Bioavailability of Trace Heavy Metals in Surface Waters — A Review, National Water Research Institute Contribution No. 84-45, Canada Centre for Inland Waters, Burlington, Ontario, Canada, 1984.
35. Leppard, G. G., Size, morphology and composition of particulates in aquatic ecosystems: solving speciation problems by correlative electron microscopy, *The Analyst,* 117, 595, 1992.
36. Krishnappan, B. G. and Ongley, E. D., River Sediments and Contaminant Transport — Changing Needs in Approach, National Water Research Institute Contribution No. 88-85, Canada Centre for Inland Waters, Burlington, Ontario, Canada, 1988.

37. Ongley, E. D., Birkholz, D. A., Carey, J. H., and Samoiloff, M. R., Is water a relevant sampling medium for toxic chemicals? An alternative environmental sensing strategy, *J. Environ. Qual.*, 17(3), 391, 1988.
38. Steen, W. C. and Karickhoff, S. W., Biosorption of hydrophobic organic pollutants by mixed microbial populations, *Chemosphere*, 10, 27, 1981.
39. Baughman, G. L. and Paris, D. F., Microbial bioconcentration of organic pollutants from aquatic systems — a critical review, *Crit. Rev. Microbiol.*, 8(3), 205, 1981.
40. Grimes, D. J. and Morrison, S. M., Bacterial bioconcentration of chlorinated hydrocarbon insecticides from aquatic systems, *Microbial Ecol.*, 2, 43, 1975.
41. Rao, S. S., Droppo, I. G., and Ongley, E., Fluvial Suspended Aggregates and Contaminant Association, National Water Research Institute Contribution No. 90–93, Canada Centre for Inland Waters, Burlington, Ontario, Canada, 1990.
42. Daniels, S. A. and Rao, S. S., Response of Aquatic Bacterial Aggregates to an Environmental Contaminant: A Compartmentalization study, National Water Research Institute Contribution No. 92–09, Canada Centre for Inland Waters, Burlington, Ontario, Canada, 1992.
43. Gschwend, P. M. and Wu, S., On the constancy of sediment water partition coefficients of hydrophobic organic pollutants, *Environ. Sci. Technol.*, 19, 1985.

INDEX

D

E

F

G

H

I

K

L

M

N

O

P

Q

R

S

T

U